A PRACTICAL FIELD GUIDE FOR ISO 13485:2016

A PRACTICAL FIELD GUIDE FOR ISO 13485:2016

Medical Devices—Quality Management Systems—Requirements for Regulatory Purposes

- Leadership Guidance
- Implementation Support
- Documentation Assistance
- Auditing Technique
- Revisions/Updates
- Senior Management Briefs

Erik V. Myhrberg and Joseph Raciti
with Brandon L. Myhrberg

Quality Press
Milwaukee, Wisconsin

American Society for Quality, Quality Press, Milwaukee 53203

Printed in the United States of America
25 24 23 LS 5 4

Publisher's Cataloging-in-Publication Data

Names: Myhrberg, Erik V., author. | Raciti, Joseph, author. | Myhrberg, Brandon L., author.
Title: A Practical field guide for ISO 13485:2016 : medical devices—quality management systems—requirements for regulatory purposes / Erik V. Myhrberg and Joseph Raciti ; with Brandon Myhrberg.
Description: Includes bibliographical references. | Milwaukee, WI: Quality Press, 2019.
Identifiers: LCCN 2019951103 | ISBN 978-0-87389-990-1 (spiral) | 978-1-951058-21-0 (epub) | 978-1-951058-22-7 (pdf)
Subjects: LCSH Equipment Design—standards—United States. | Equipment and Supplies—standards—United States. | Industry—standards—United States. | Quality Control—United States. | BISAC MEDICAL / Instruments & Supplies
Classification: LCC R856.A2 M95 2019 | DDC 610.28—dc23

R856.15.M94 2012
610.28'4—dc23 2012039034

ISBN: 978-1-63694-132-5 (Paperback)
ISBN: 978-0-87389-990-1 (Spiralbound)
ISBN: 978-1-951058-21-0 (EPUB)
ISBN: 978-1-951058-22-7 (PDF)

ASQ advances individual, organizational, and community excellence worldwide through learning, quality improvement, and knowledge exchange.

Bookstores, wholesalers, schools, libraries, and organizations: Quality Press books are available at quantity discounts with bulk purchases for business, trade, or educational uses. For more information, please contact Quality Press at 800-248-1946 or books@asq.org.

To place orders or to browse the full selection of Quality Press titles, call 800-248-1946 or visit our website at: http://www.asq.org/quality-press.

Quality Press
600 N. Plankinton Ave.
Milwaukee, WI 53203-2914
Email: books@asq.org
ASQ Excellence Through Quality™

To our families, colleagues, and fellow professionals.

TABLE OF CONTENTS

INTRODUCTION

Almost 30 years have passed (and over 150 international projects) since we first became aware of, and started using, the ISO standards. So much has changed and evolved during this time, including the advent of the ISO 14000 series of environmental management standards, the addition and deletion of standards within the ISO 9000 series, and the creation of a host of industrial and sector-specific standards and reports based on ISO 13485. In addition, a natural extension of the standard has been prepared for the management of quality systems for the medical and medical device sector.

Even with all this progress, the fundamental use of the ISO 9000 series (up to and including ISO 13485) for the medical and medical device sector has not changed. Companies, teams, and individuals are still trying to meet customer expectations. Worldwide competition still drives the need for innovation. And internal process pressures still demand continual improvement in order to remain functional and vital as an organization.

This field guide has been created in order to foster an inter-reliance between senior management, middle management, functional teams, and the individual. Users of the field guide will find practical tools, tips, and techniques useful for developing, implementing, and maintaining a quality management system.

It is our hope that the readers and users of this field guide will benefit from our years of extensive field and classroom experience in the application of any ISO 9000 quality management system series standard for the medical and medical device sector.

As we forge into the twenty-first century, companies and organizations of all sizes are being pushed to the limit of their resources. At one time it was sufficient to meet most of our customers' requirements. Today, on a global scale, we are being asked to do more with less, for less, without compromising quality or delivery. At some point soon, the current internal systems will not be able to hold back the deluge, and companies will be faced with a stark decision—consistently improve or perish.

One of the best and most widely accepted ways in which companies can face these challenges is to implement an effective *and* efficient quality management system, which not only adds value to the organization but also satisfies the customer. Indeed, this is already the century of international standards. The ISO 13485:2016 series of standards is both useful to the organization and here to stay.

Use this guide diligently as you work through your ongoing ISO efforts!

Erik Myhrberg, Joe Raciti, and Brandon Myhrberg
http://www.moorhill.com

HOW TO USE THIS FIELD GUIDE

The purpose of this field guide is to assist organizations, step by step, in implementing a *quality management system* (QMS) for the medical and medical device sector in conformance with ISO 13485:2016, whether from scratch or by transitioning from variations of the ISO 13485 family. In keeping with ISO 9000:2015's definition of quality as "the degree to which a set of inherent characteristics fulfills requirements," we have identified the requirements and inherent characteristics, or distinguishing features, for this field guide. Within the guide, each subclause containing requirements is the focus of a two-page spread that consistently highlights features that fulfill the requirements listed below.

WHAT THE FIELD GUIDE WILL DO FOR YOU

- Provide a user-friendly guide to ISO 13485:2016 requirements for implementation purposes.
- Identify the documents/documentation required, along with recommendations about what to retain or add to a QMS during ISO 13485:2016 implementation.
- Guide internal auditor(s) regarding questions that verify a conforming and effective QMS exists and is being maintained.
- Direct management on ways to satisfy ISO 13485:2016 enhanced requirements, as well as responsibilities for top management.
- Depict step-by-step what must occur to create an effective, conforming QMS.

INHERENT CHARACTERISTICS, OR WHAT THE FIELD GUIDE PROVIDES

- *The standard.* A paraphrase of what a subclause of ISO 13485:2016 requires in easy-to-understand language, with references to information in ISO 9000:2015 and guidance in ISO 9004:2015 to enhance the user's understanding of what ISO 13485:2016 requires and what possible added steps the user can take to improve performance.
- *Document requirements.* A list of the documentation/documents required by the subclause, with ideas to consider for satisfying those requirements that will go beyond meeting them and move toward continual improvement.
- *Internal audit questions.* What every internal audit team needs to ask, at a minimum, when assessing the QMS for conformance with the subclause.
- *Management summary.* A concise description of what management must do and/or is responsible for in order to achieve conformance to the subclause, along with some guidance on additional steps management can take to enhance the overall QMS.
- *Subclause flowchart.* A depiction of the necessary steps during an implementation/transition effort to effectively and efficiently satisfy the requirements of the subclause of ISO 13485:2016, along with a tip box providing guidance on use of the flowchart.

This medical device field guide is designed to provide you with a consistent approach to implementing an ISO 13485:2016–conforming QMS, which is appropriate since ISO 13485:2016 continues to view quality as the ability of an organization to consistently deliver product that meets customer requirements.

The field guide examines each subclause of Sections 4–8 of ISO 13485:2016, which contain the requirements, with characteristics 1–4 presented on one page and characteristic 5 presented on the facing page.

The example on the following pages presents subclause 1, Scope, which does not contain requirements but is critical for properly excluding (only from Sections 6, 7, or 8) any requirements of ISO 13485:2016 that do not apply to your organization's QMS. Consider this an important method for establishing the QMS's scope.

What separates this field guide from most other books on ISO 13485:2016 and its implementation are the flowcharts showing the steps for implementing a QMS that meets a subclause's requirements.

The flowcharts themselves can be overwhelming when you first look at them. For this reason, an information box appears with each flowchart that explains pertinent facts and/or what the flowchart represents and how it is to be used.

SUMMARY OF EACH SECTION

Each section ends with a summary, an outlined way of looking at the standard. Reading these sections of the field guide will not only help prepare your company for a new implementation of the medical device standard but will also aid those companies already involved with the standard in drastically improving the current management system by providing specific points to consider.

ISO 13485 STRATEGIC OVERVIEW

In order to truly make the most of your management system, presented below is a list of the major reasons for the standard as well as the changes to the standard:

- Independent standard for medical device organizations
- First edition released in 1996
- Second edition released in 2003
- Latest edition released in 2016
- Current edition prepared by technical committee ISO/TC 210, Quality management and corresponding general aspects for medical devices
- Developed with ISO 9001 in mind
- Stand-alone quality standard, not linked to ISO 9001
- ISO 13485:2016 and ISO 9001 are mutually exclusive (that is, certification to one does not imply certification to the other)
- Focus on how processes work together to improve the business
- Focus on activities that follow a trigger event and the interactions between those activities
- Not bound or defined by procedures or departments
- Moved from a procedure-based approach to a process-based approach

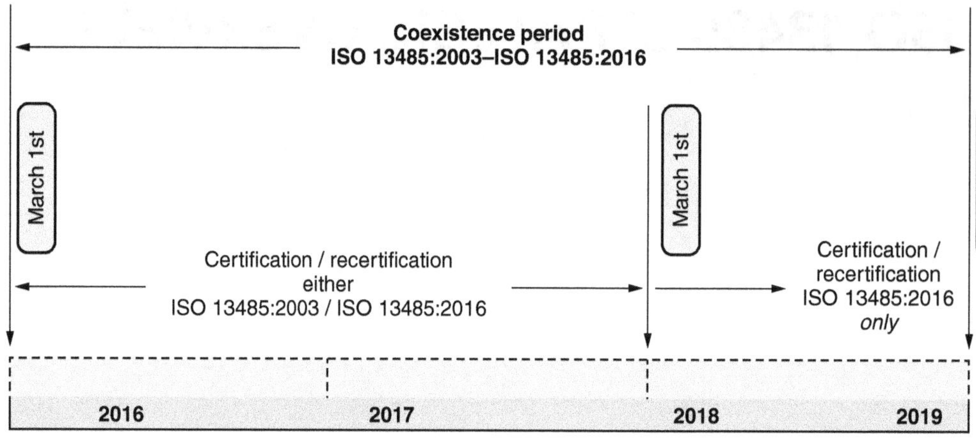

- Requires maintaining effectiveness of QMS, not continual improvement of QMS.
- Focus on meeting customer requirements, not achieving customer satisfaction.
- Within ISO 13485:2016 there are two appendices:
  - Appendix A: shows the changes from the 2003 version to the current 2016 version.
  - Appendix B: illustrates the differences between the 13485 and 9001 versions.

0.1 General

- Requirements for a quality management system (QMS) used by organizations for design, development, production, installation, and servicing of medical devices, including related services; requirements can also be applied by suppliers and external parties providing products or services to these organizations

0.2 Clarification of concepts

- Understanding of the terms and concepts applied throughout the standard

0.3 Process approach

- Basis for this standard; any activity that receives inputs and converts them to outputs is a process

0.4 Relationship with ISO 9001

- While this is a stand-alone standard, it's based on ISO 9001

0.5 Compatibility with other management systems

- Standard does not include requirements specific to other management systems

1.1 Requirements of a QMS for organizations providing medical devices and related services for 2016 that consistently meet customer requirements and applicable regulatory requirements

1.1 Facilitates harmonized medical device regulatory requirements

1.1 Includes some medical device requirements of ISO 9001 and excludes some requirements not appropriate as regulatory requirements

1.1 Organizations whose QMS conforms to this standard cannot claim conformity to ISO 9001 unless their QMS conforms to all requirements of ISO 9001

1.2 Requirements are specific to organizations providing medical devices, regardless of type or size

1.2 Regulatory requirements and/or the nature of a medical device may allow justification for exclusion *only within Clause 7*

- Reference documents indispensable for the application of the document
  - Dated (only edition cited applies)
  - Undated (latest edition of reference applies)
- ISO 9000:2015 *Quality management systems—Fundamentals and vocabulary*
- ISO/TR 14969—Technical Report intended to provide guidance for the application of ISO 13485
- Terms and definitions given in ISO 9000:2015 apply
- Definitions provided in national regulations can differ slightly and take precedence
- Key changes in definitions from ISO 13485:2003 (high level)
  - Active implantable medical device (incorporated into implantable medical device)
  - Active medical device (incorporated into medical device)
  - Authorized representative (new)
  - Clinical evaluation (new)
  - Complaint (updated)
  - Distributor (new)
  - Implantable medical device
  - Importer (new)
  - Labeling
  - Life cycle (new)
  - Manufacturer (new)
  - Medical device
  - Medical device family (new)
  - Performance evaluation (new)

– Post-market surveillance (new)

– Product (new)

– Purchased product (new)

– Risk (new)

– Risk management (new)

– Sterile barrier system (new)

– Sterile medical device

SECTION 4: QUALITY MANAGEMENT SYSTEM

The Standard: 4.0 Quality Management System

4.1 General Requirements

13485 The organization must develop, document, set up, and maintain a QMS and continually improve its effectiveness through six activities, including identification of needed processes, making sure adequate resources and information are available to support both the processes and their monitoring, and achieving expected results and continual improvement of the processes though implemented actions. All related activities must conform to ISO 13485's requirements. When outsourcing related to product occurs, the QMS must contain provisions to ensure that the organization controls the processes involved. A *Note* indicates processes that should be included in the QMS.

9004 Offers guidance on activities top management should pursue to create a customer-oriented organization and move it and its QMS toward continual improvement and improved performance. Annexes A and B provide guidance on self-assessment and process improvement.

Document Requirements:

Required:

- No specific documents, although this subclause relates to and affects other requirements in ISO 13485:2016 that concern specific QMS documents/documentation.
- If the organization uses computer software for the application of the QMS, there must be documented procedures and associated records for IT validation.

Remember:

- Processes need to be clearly defined, with the sequence and interactions between stages and with other processes defined. This will require documenting the organization's operations.
- The defined processes must include all administration and management activities.

Internal Audit Questions:

- Has the QMS been established, documented, implemented, and maintained?
- Have processes, including sequences and interactions that are necessary for maintaining the QMS, been determined?
- Has top management clearly communicated the QMS throughout the organization?
- Has the organization demonstrated continual improvement?
- Does the organization have control over processes that are outsourced?
- Does the organization utilize computer software applications in the QMS?

Management Summary:

Top management must:

- Define systems and processes.
- Ensure that they are understood throughout the organization and managed.
- Ensure effective control of processes.
- Utilize data to evaluate performance.
- Identify the use of computer software applications and validate them.

Top management must follow these QMS principles:

- Have a true customer focus.
- Demonstrate leadership.
- Involve people.
- Use a process-oriented approach.
- Use a systems approach to management.
- Strive for continual improvement.
- Base decisions on facts and data.
- Cultivate mutually beneficial supplier relationships.

4.1 General Requirements

The organization shall:

- Establish
- Document
- Implement
- Maintain
- Continually improve

a quality management system.

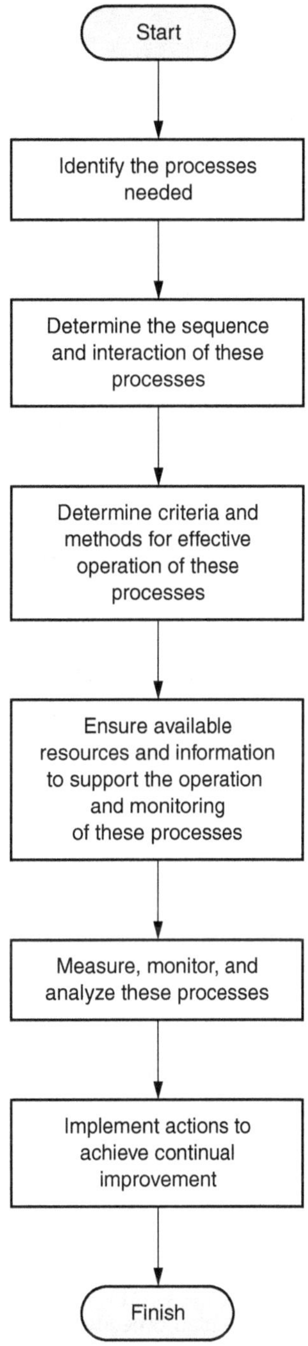

The Standard: 4.0 Quality Management System

4.2 Documentation Requirements

4.2.1 General

13485	Specifies what is to be included in the QMS documentation: quality policy, objectives, and manual; procedures that must be documented to conform to ISO 13485; documents required for "effective planning, operation and control" of the organization's processes; and quality records. Three *summaries* define what is meant by "documented procedure," clarify that the amount and level of QMS documentation will vary between organizations for three reasons, and explain that documentation can be in any form or medium (for example, flowchart, electronic file).
9004	Suggests that management should define the required QMS documentation for the organization, explains what the documentation should do, and provides six factors management should consider in defining the QMS documentation.
9000	Defines a *document* in 3.7.2 as "information and its supporting medium" and provides examples of types of documents. Explores in 2.7, Documentation, the value of documentation and the types of documents used in a QMS.

4.2.2 Quality Manual

13485	A quality manual is to be developed and maintained that includes the QMS's scope—with full explanation for why any requirements are excluded—its documented procedures (or indications of where to find them), and a description of the process approach as it works in the QMS.
9000	Defines a *quality manual* in 3.7.4 as a "document specifying the [QMS] of an organization."

Document Requirements:

Remember:

- A quality policy (statement *and* objectives), quality manual, medical device file, and the requisite procedures specified document your organizational needs to function and improve while meeting customer/regulatory requirements and record conformance.
- The notes in this subclause section can have a significant bearing on the documentation when establishing the QMS.

Internal Audit Questions:

- Does the QMS include documented statements of the *quality policy* and *quality objectives*?
- Is there a *quality manual*?
- Are there other national or regional regulations that apply?
- Are the required documented procedures in place?
- Are there additional documented procedures where determined as needed by the organization?
- Is the quality management system documentation based on the size, type, complexity, and interaction of processes in the organization?
- Are the required records available?

Management Summary:

Management must define the documentation, including records that:

- Are required to establish, implement, and maintain a QMS.
- Support an effective and efficient operation.

4.2 Documentation Requirements

4.2.1 General

4.2.2 Quality Manual

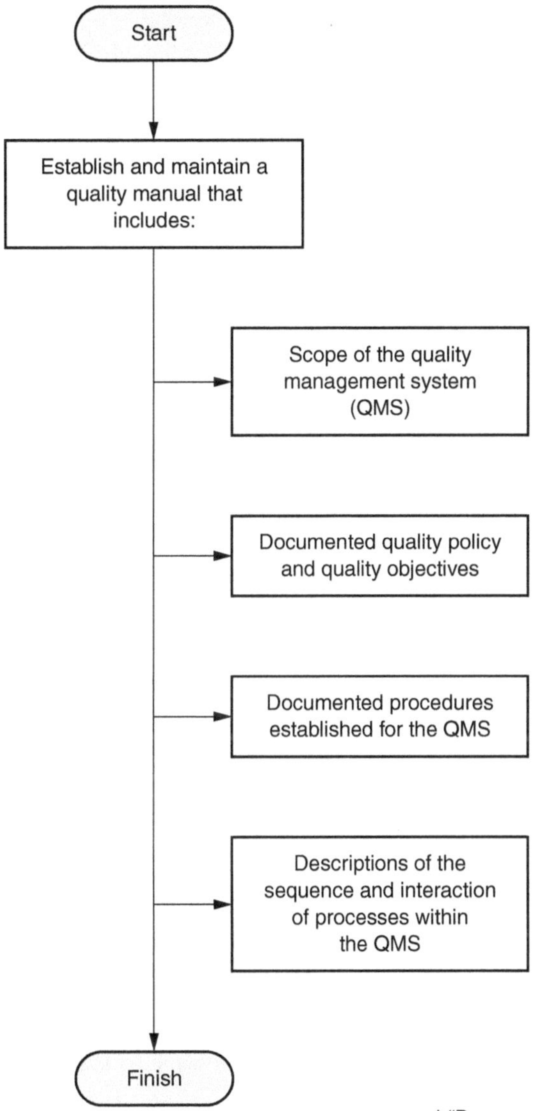

* "Documented procedures" must be:

- Established
- Documented
- Implemented
- Maintained

The Standard: 4.0 Quality Management System

4.2 Documentation Requirements

4.2.3 Medical Device File

13485	Medical device file plus a detailed list of items (a–f) that shall be included to meet regulatory requirements.

4.2.4 Control of Documents

13485	The organization must develop and document a procedure that defines the necessary controls to ensure that eight document-related processes are established and provide the level of document control required by ISO 13485. These processes include document approval, review, updating and reapproval, revision status identification, availability of documents where they are needed, legibility, identification and control of external documents, and prevention of unintended use of obsolete documents. The retention period for documents specific to the manufacture and testing of medical devices must be defined. The minimum retention period is not to be less than the life of the device as defined by the organization or relevant regulatory requirement.
9004	Advises that control of documentation should be evaluated to ensure that the procedures support the organization's effectiveness and efficiency when the eight criteria are considered.

Document Requirements:

Required:

Document control procedure

Remember:

The review, approval, updating, and reapproval process has always been a requirement; however, you must now add the concept that the documents themselves must be reviewed.

Internal Audit Questions:

- Is document control established in a documented procedure?
- Are documents reviewed, updated, and approved for adequacy prior to use? Are document changes reapproved to ensure adequacy prior to use? Is current document revision status maintained?
- Are obsolete documents that are retained for any purpose suitably identified to prevent unintended use?
- Is there a process to ensure that documents remain legible, readily identifiable, and retrievable?
- For medical devices, are there specific retention requirements for document/record retention?
- Are versions of applicable documents available at points of use?
- Is there a process for maintaining control and currency of "point of use" documents?
- Are external documents identified and controlled?

Management Summary:

Management should:

- Define the documentation required.
- Give consideration to and, where practicable, describe how to address contractual requirements, acceptance of standards (for example, medical device–specific requirements), regulatory requirements, organizational decisions, and the needs and expectations of interested parties.
- Evaluate the generation, use, and control of documentation against criteria such as functionality, user-friendliness, resources, policies and objectives, managing knowledge, benchmarking documentation, and interfaces used by outside sources.

4.2 Documentation Requirements (continued)

4.2.3 Medical Device File

4.2.4 Control of Documents

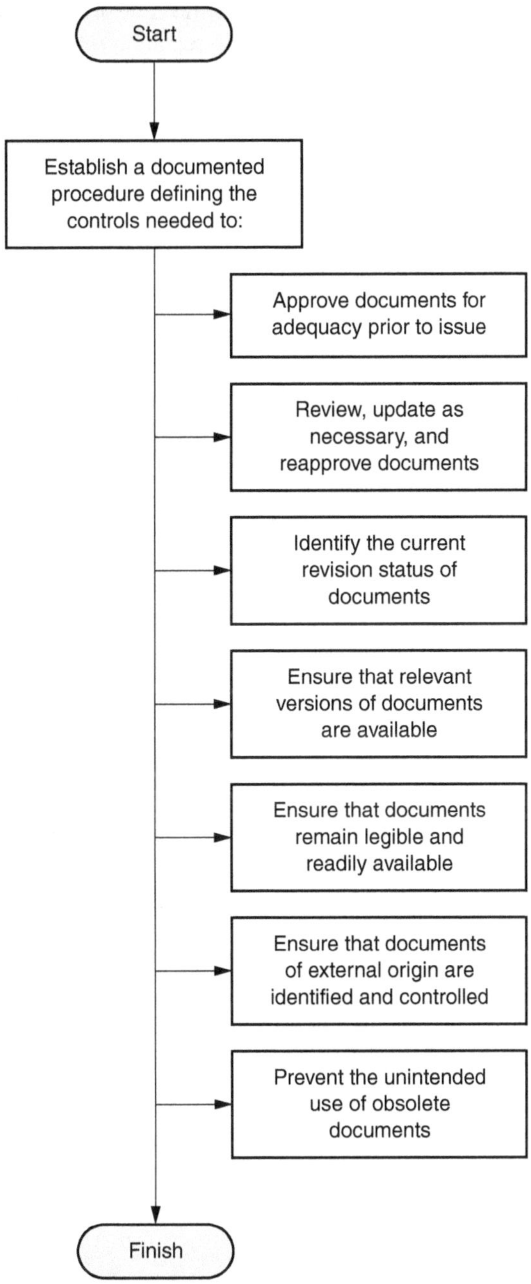

The Standard: 4.0 Quality Management System

4.2 Documentation Requirements

4.2.5 Control of Records

13485	Quality records are to be "established and maintained" to prove that the QMS is operating as intended and is effective and satisfies the requirements of ISO 13485. The organization must establish a specific procedure in addition to those under 4.2.3 to define the controls that will apply to quality records and specify, among other things, their storage, retrieval, retention time, and disposition.
9000	Defines a *record* in 3.7.6 as a "document stating results achieved or providing evidence of activities performed."

Document Requirements:

Required:

Documented procedure

Remember:

The procedure must not only control records but do so in a way that serves the organization's needs without creating valueless paperwork and bureaucracy.

Internal Audit Questions:

- Is there a documented procedure for control of records?
- Have the organization's records been identified?
- Are protection requirements in place for all records?
- Have retention times and disposition requirements been established for all records?
- Are records disposed of as required by the established procedures?
- Have storage and retrieval requirements been determined and implemented for all records?

Management Summary:

- Management must establish and maintain records in order to provide objective evidence of conformance to the organization's QMS.
- Give consideration to contractual requirements, acceptance of standards (for example, medical device–specific requirements), time frames for potential feedback (for example, the lifetime of a device), and the needs and expectations of interested parties (for example, customers and authorities).

4.2 Documentation Requirements (continued)

4.2.5 Control of Records

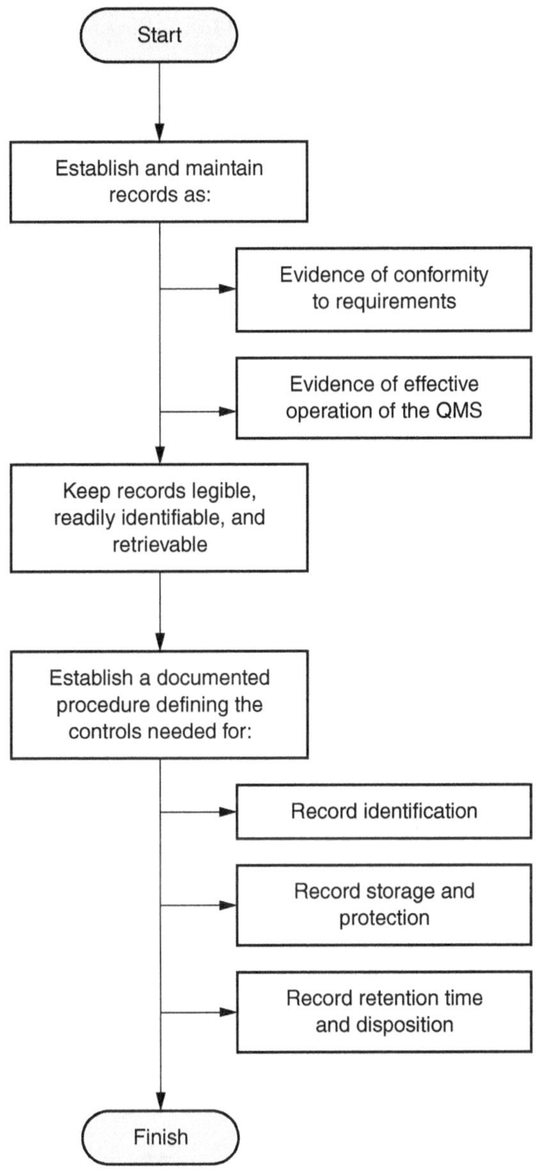

Clause	Change	Interpretation
0.4 (Relationship with ISO 9001)	Reference to ISO/TR14969 has been removed since there has been no decision to revise this standard.	No interpretation required since this is a deletion from the 2003 version of this standard.
0.5 (Compatibility with other management systems)	The reference to ISO 13485, following the format of ISO 9001, has been removed.	No interpretation required since this is a deletion from the 2003 version of this standard.
4.1.1 (General requirements)	Inclusion of documenting the roles of the organization and determination of processes.	Roles within the organization now need to be defined, for example, manufacturer, distributor, importer.
4.1.2 (General requirements)	Application of a risk-based approach.	Organizations are now required to apply a risk-based approach to control processes within the QMS.
4.1.3 (General requirements)	Organizations required to be in compliance with regulatory requirements.	Organizations are required to comply with applicable regulatory requirements, for example, national or regional.
4.1.4 (General requirements)	Changes to processes shall be assessed as to their impact on the QMS.	Requires organizations to document the assessment of changes to QMS, product, and process.
4.1.5 (General requirements)	Organizations are required to demonstrate compliance/control of outsourced processes with applicable regulatory requirements, based on risk.	Compliance/control can be demonstrated via use of a quality agreement between the organization and its outsourced service provider.
4.1.6 (General requirements)	Documentation of software validation, with satisfactory completion, review, and approval.	Requirement to document all software validations based on risk posed by software and its application to the quality management system.
4.2.1 (General)	Notes 1 and 2, which originally provided additional information relating to general requirements, have been removed.	The removed Notes from the 2003 version related to the variation of QMS documentation between organizations, and that documentation could be in any form or type of medium.

SUMMARY

by S. Pletzer

ISO 13485:2016—4 Quality Management System

4.1 General requirements

- Identify processes needed and their application
- Determine sequence and interactions of these processes
- Ensure that operation and control of these processes is effective
- Ensure availability of resources
- Monitor, measure, and analyze these processes
- Implement actions to achieve results and maintain process effectiveness
- Outsourced processes must also be included

4.2 Documentation requirements

- Quality policy, quality objectives, and quality manual
- Procedures, records, or others as required
- Control of documents
- Control of records

SECTION 5: MANAGEMENT RESPONSIBILITY

5.1 Management commitment
5.2 Customer focus
5.3 Quality policy
5.4 Planning
  5.4.1 Quality objectives
  5.4.2 Quality management system planning
5.5 Responsibility, authority and communication
  5.5.1 Responsibility and authority
  5.5.2 Management representative
  5.5.3 Internal communication
5.6 Management review
  5.6.1 General
  5.6.2 Review input
  5.6.3 Review output

The Standard: 5.0 Management Responsibility

5.1 Management Commitment

13485 Top management must be committed to creation and implementation of a QMS and its continual improvement through four types of activities. These activities include organization-wide communication of the criticality of meeting customer and other requirements, establishment of the quality policy and objectives, management reviews, and making sure QMS resources are available.

9004 Suggests actions top management should consider to maximize customer satisfaction and achieve benefits for all parties, recommends methods of performance measurement to help achieve objectives, explores activities relating to the quality management principles that top management should "demonstrate leadership in, and commitment to," and notes activities that should be considered to optimize processes.

9000 Discusses what management can do in 2.6, Role of top management within the quality management system.

Document Requirements:

Required:

No specific documents, although evidence of top management's commitment to the QMS must be present.

Remember:

- This clause is a reminder to top management that there must be a process not only to create awareness of the organization's quality policy and quality objectives, but also to *maintain* this awareness.
- In addition, top management needs to be aware of the enhanced responsibilities and roles it has with ISO 13485:2016 and the need to demonstrate its commitment to the QMS in ways that can be not just audited by auditors but demonstrated by the rest of the organization to maintain and improve processes.

Internal Audit Questions:

- Has management established a quality policy and developed quality objectives to support the policy?
- Are management reviews performed?
- Do any other review activities exist?
- Does management provide and review the adequacy of resources?
- Is a process in place to ensure that all employees understand the importance of fulfilling customer, regulatory, and legal requirements?
- Is a process in place to develop specific programs for communicating customer and regulatory requirements?
- Is there objective evidence of conformance to these programs in review records?

Management Summary:

- Top management should be a model of leadership, show commitment, and be actively involved in achieving customer satisfaction as well as regulatory and legal compliance.
- Top management must establish the quality policy as a sign of its commitment to the organization's mission/vision, as well as the continual improvement of the QMS and its effectiveness.
- Top management must define measurement methods to determine whether planned objectives have been achieved.
- When developing, implementing, and managing the QMS, top management must be involved in the development of quality objectives, either by participating in their setting or by providing the framework and environment for their creation.
- Top management must evaluate the QMS regularly.
- When developing, implementing, and managing the QMS, top management must consider the principles listed in the Management Summary section for clause 4.1.

5.1 Management Commitment

These are the actions required of the organization's top management.

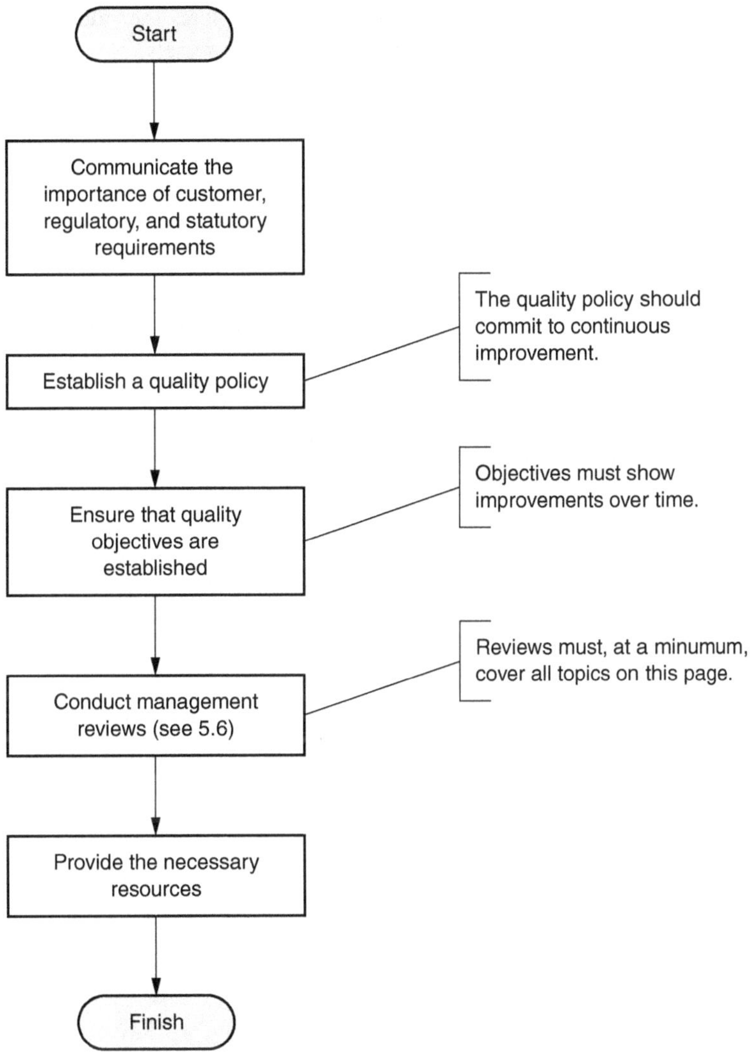

The Standard: 5.0 Management Responsibility

5.2 Customer Focus

13485	To increase customer satisfaction, top management must ensure that the organization understands and is capable of satisfying customer requirements (see 7.2.1 and 8.2.1).
9004	Provides examples of customer/end user needs, and guidance on what an organization should do to understand and satisfy the needs of all interested parties. Explores the value of partnerships and what an organization should undertake regarding its relationship with society.

Document Requirements:

Remember:

Top management must be able to demonstrate that they have put in place processes to make certain that these requirements are met.

Internal Audit Questions:

- Is there a process in place for meeting customer needs and expectations?
- Are there processes in place to convert customer needs and expectations into company-specific requirements?
- What objective evidence is available to confirm that these processes are utilized?
- How has top management led the organization in such a way that the organization's focus is on meeting the needs of the customers, not just making a product?

Management Summary:

- The organization should identify personnel needs and expectations for recognition, work satisfaction, and personal development in order to ensure the involvement and motivation of its people in relation to customer satisfaction.
- In order to understand and satisfy the current and future needs of customers, the organization should:
  - Identify, understand, and respond to customer needs and expectations.
  - Translate customer needs and expectations into requirements.
  - Communicate the requirements throughout the organization.
  - Focus on process improvements to ensure satisfaction.

5.2 Customer Focus

Top management must demonstrate a strong commitment to customer satisfaction.

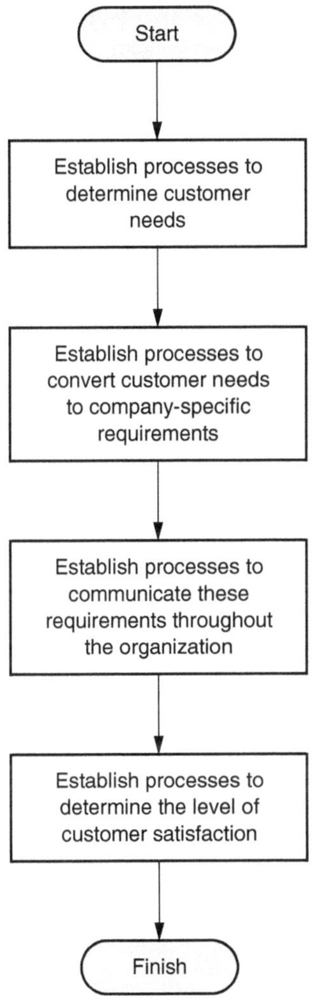

The Standard: 5.0 Management Responsibility

5.3 Quality Policy

13485	Top management is responsible for guaranteeing that the quality policy is appropriate and commits the organization to conformance with applicable requirements and to continual improvement of the QMS, and that it is a suitable basis for establishing quality objectives, is disseminated and understandable to the organization, and is reviewed for ongoing appropriateness to the organization.
9004	Lists what top management should consider in establishing the quality policy and spells out conditions that will allow it to be used for QMS improvement.
9000	Defines a *quality policy* in 3.2.4 as a formal expression of the "overall intentions and direction of an organization related to quality . . . ." 2.5, Quality policy and quality objectives, explains the policy's purpose and role and its relationship with the objectives.

Document Requirements:

Required:

Quality policy statement

Remember:

The quality policy will now need to include commitment to continuous improvement and to measurable objectives.

Internal Audit Questions:

- Is the quality policy documented?
- Does the quality policy provide a framework for establishing and reviewing the quality objectives?
- Does the quality policy demonstrate a commitment to meeting all relevant requirements and include provisions for continual improvement?
- Does management review the quality policy regularly for continuing suitability? Do the management review records indicate that the quality policy has been reviewed regularly for suitability?
- Do the quality objectives provide measurable evidence that the organization is achieving their stated quality policy?
- Are the employees of the organization clear on what the policy means and understand their role in carrying out the policy?

Management Summary:

- Top management is responsible for setting this overarching vision for the QMS, from which the objectives will cascade. Top management must utilize the quality policy as a means of leading the organization toward improvement of its performance.
- The policy must be ambitious but realistic and relevant to the organization.

5.3 Quality Policy

Top management
must establish *and* maintain
an appropriate quality policy.

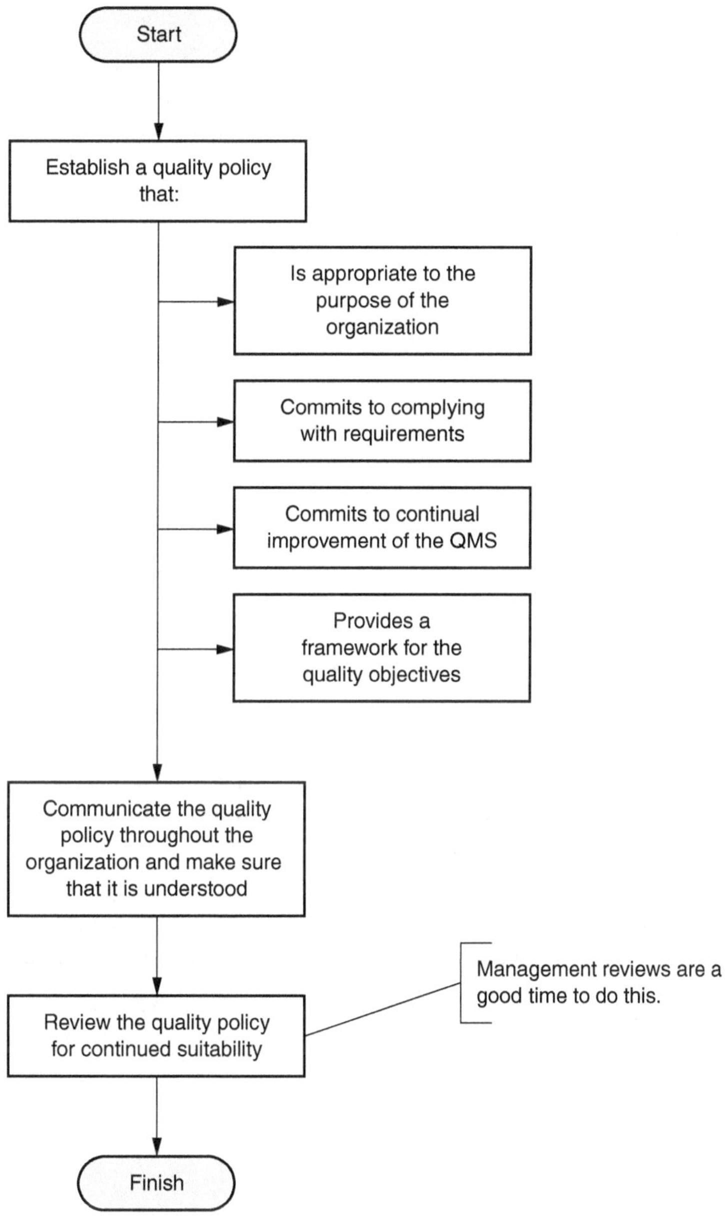

The Standard: 5.0 Management Responsibility

5.4 Planning

5.4.1 Quality Objectives

13485	Consistent, measurable quality objectives must be set under top management's auspices wherever they are required within the organization to satisfy product requirements, and be based on the foundation provided in the quality policy.
9004	Indicates that quality objectives lead to improved organizational performance and recommends that top management establish them based on the organization's strategic planning and quality policy. Suggests factors worth considering in setting objectives and then communicating the objectives so everyone contributes to their achievement.
9000	Defines *quality objective* in 3.2.5 and explains in 2.5, Quality policy and quality objectives, their purpose/role within the policy and the benefits of achieving the objectives.

Document Requirements:

Required:

Quality policy objectives

Remember:

The need for objectives to be quantifiable in form as actual input on QMS effectiveness, which was only implied in ISO 13485:2003, is now a specific requirement.

Internal Audit Questions:

- Have measurable quality objectives been established at relevant functions and levels within the organization? How are these documented?
- Do quality objectives include a method of determining and documenting whether processes are meeting requirements for the organization's and the customers' products and/or services?
- Has the organization identified the resources, activities, and processes required to meet the quality objectives (quality management system, product and/or service realization, and verification processes)?

Management Summary:

- The organization should use its strategic planning process and quality policy as a framework for setting quality objectives.
- Top management sets quality objectives as a method for leading continual improvement within the organization.
- Quality objectives must be measurable and communicated in such a way that all personnel can contribute to their fulfillment.
- Quality objectives must be regularly and systematically reviewed and revised.

5.4 Planning

5.4.1 Quality Objectives

Top management
must ensure that appropriate
quality objectives
are established.

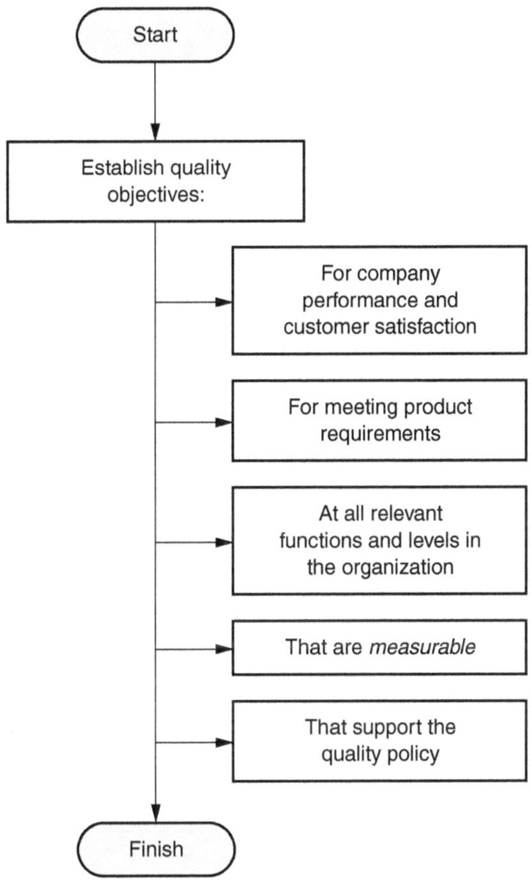

The Standard: 5.0 Management Responsibility

5.4 Planning

5.4.2 Quality Management System Planning

13485	Top management must ensure that planning relative to the QMS is conducted so as to conform with the requirements of 4.1, General requirements, and achieve the organization's quality objectives. This includes ensuring that QMS revisions do not compromise ISO 13485 conformance and effectiveness of the QMS.
9004	Discusses the broader concept of quality planning and the inputs and outputs that could make the QMS and quality planning effective and efficient in meeting an organization's quality objectives and strategic requirements.
9000	Defines *quality planning* in 3.2.9 as management activities "focused on setting quality objectives and specifying necessary operational processes and related resources to fulfill the quality objectives."

Document Requirements:

Requirements:

A documented plan is required that identifies the resources needed to achieve quality objectives.

Remember:

Changes are to be introduced without failures of the QMS.

Internal Audit Questions:

- Has the organization identified and planned processes and resources for the quality system?
- Do the processes provide for the control of changes to the quality management system?
- As organizational changes occur, does quality planning take into account the needs of the organization?
- What objective evidence exists to demonstrate that quality planning contributes to continual improvement (that is, program briefs, action items in a management review record, or references to an improvement in a procedure)?

Management Summary:

- Top management must establish and authorize quality plans for the organization.
- Top management should plan and monitor processes to effectively and efficiently fulfill the organization's quality objectives.
- Top management must evaluate changes to quality plans to ensure that they are not in conflict with the requirements.

5.4 Planning

5.4.2 Quality Management System Planning

Top management is responsible for both the planning and the integrity of the quality management system.

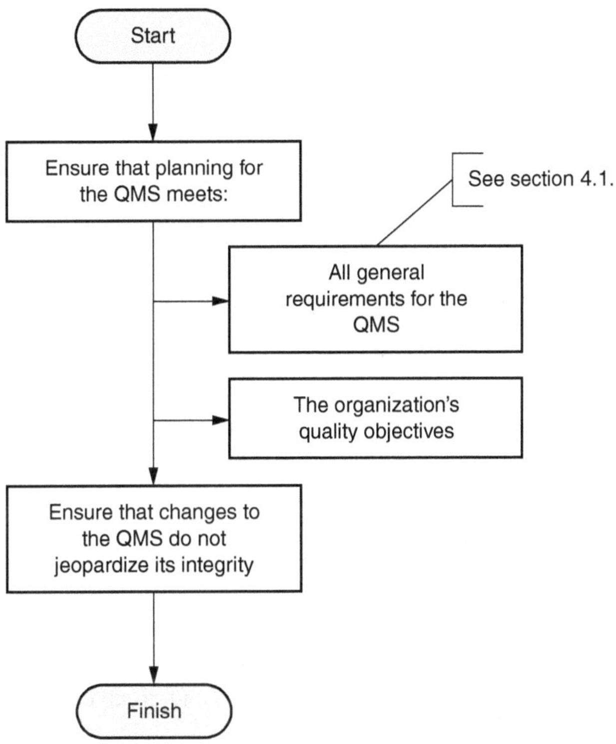

The Standard: 5.0 Management Responsibility

5.5 Responsibility, Authority and Communication

5.5.1 Responsibility and Authority

13485	The responsibilities and authorities for management of an organization's QMS and the interrelationships between positions (jobs) are to be defined and communicated throughout the organization, with top management responsible for guaranteeing this is accomplished.
9004	Recommends that top management should define and communicate the responsibilities and authorities, with a range of employees empowered to realize their contributions to quality objective fulfillment and their buy-in to the QMS.

Document Requirements:

Remember:

- Organizational chart
- Job descriptions
- Other documents defining the authority and responsibility of each job function

Internal Audit Questions:

- Are responsibilities and authorities defined and communicated to facilitate effective quality management?
- Are the quality objectives consistent with the stated quality policy, and do they provide measurable evidence that the organization is achieving the policy (statement and objectives)?
- Are there requirements for authorization of key personnel (such as in reporting of adverse events, and so on)?
- Do job descriptions for key personnel include the proper authority for effective job performance?

Management Summary:

- The processes within the organization must enable people to contribute to the achievement of the quality objectives, thus better aligning the business with the management system.
- Not all quality objectives will be suitable to and attainable by all operations or personnel, so top management must provide the direction and processes by which objectives can be set for each area and group of personnel.
- Job descriptions must be developed with the proper authority for a given function to effectively apply the quality objectives.

5.5 Responsibility, Authority and Communication

5.5.1 Responsibility and Authority

Top management
must see that responsibilities
and authorities are defined and
are communicated throughout
the organization.

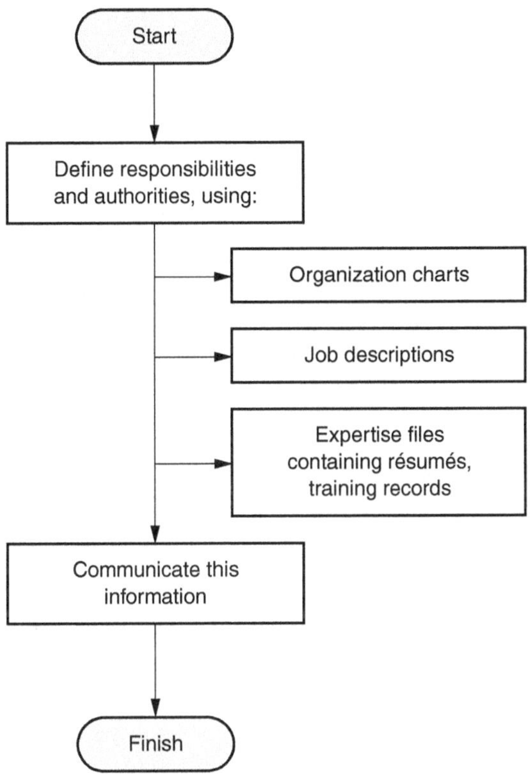

The Standard: 5.0 Management Responsibility

5.5 Responsibility, Authority and Communication

5.5.2 Management Representative

13485	Top management must appoint a management-level representative who is to ensure that a QMS is put in place and kept in operation, report to top management on how well the QMS is functioning and the actions required to correct/improve the QMS, and ensure that customer and regulatory requirements are communicated to and understood by all affected employees (that is, listening to the voice of the customer). The representative can serve as a liaison to external parties on QMS issues.
9004	Advocates having the representative play the most active role in managing and overseeing the QMS and communicating with customers and others about the QMS.

Document Requirements:

Remember:

- No documents/documentation are required, although a documented QMS planning process and/or records of planning activities are recommended
- Though not for creation, implementation, and revision of the QMS, a documented process for QMS planning and review activities will assist those involved in those planning activities and will help the organization, thus making a documented procedure and records of the QMS planning process useful.

Internal Audit Questions:

- Has the organization appointed one or more management representatives? Are the responsibilities and authorities of the management representative defined?
- As organizational changes occur, does QMS planning take into account the needs of the organization?
- Has the management representative implemented a system/program for communicating customer and regulatory requirements to the organization?
- Is there objective evidence to demonstrate that this communication has taken place?
- Does the management representative report to top management on the performance of the QMS?
- Is there objective evidence to demonstrate that this communication has taken place?
- Does the management representative promote awareness of customer and regulatory requirements throughout the organization?

Management Summary:

- The role of the management representative is to enhance effective and efficient operation and improvement of the QMS.
- The management representative must be authorized by top management.
- The management representative has the additional responsibility to promote awareness of customer and regulatory requirements throughout the organization.
- Ensure that planning takes place to produce and update an ISO 13485:2016–conforming QMS that meets the needs of the organization.

5.5 Responsibility, Authority and Communication

5.5.2 Management Representative

Top management
must appoint a manager
authorized to oversee the
quality management system.

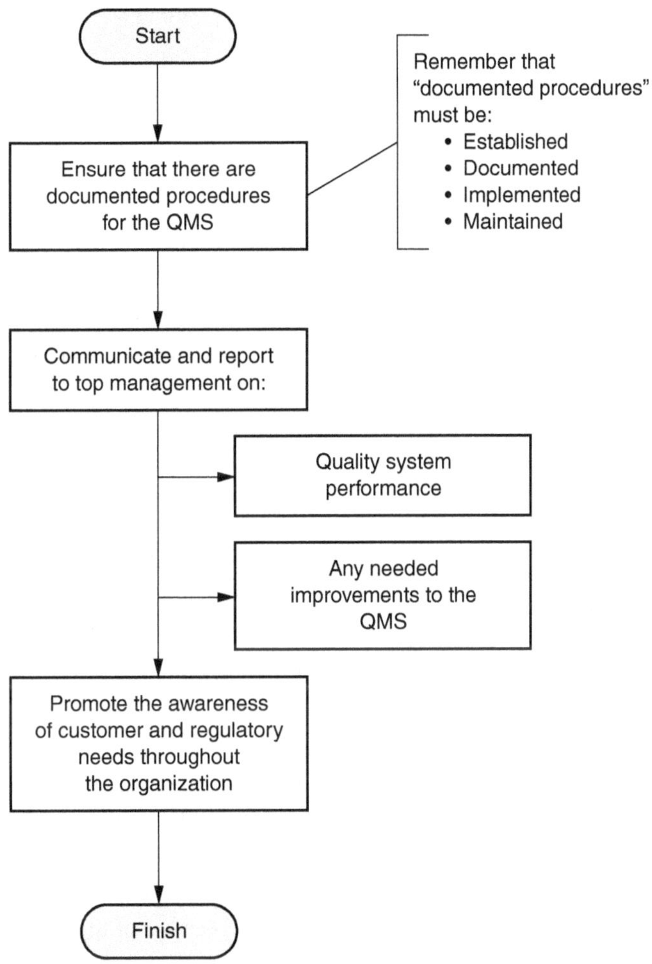

The Standard: 5.0 Management Responsibility	
5.5 Responsibility, Authority and Communication	
5.5.3 Internal Communication	
13485	Top management must ensure that the organization adopts communication processes that inform all within the organization about how effective the QMS is.
9004	Recommends that management take direct responsibility for installing communication processes and including information on certain QMS elements and accomplishments to encourage employee participation in performance improvements, achievement of quality objectives, and provision of QMS feedback. Lists ways to communicate with employees.

Document Requirements:

Remember:

No documents/documentation are specified, although documentation is the most effective means of defining and communicating QMS responsibilities and authorities. The following types of documentation can be used to define and communicate: documented procedures and records (minutes, and so on).

Internal Audit Questions:	*Management Summary:*
• What audit evidence exists to demonstrate effective internal communication (that is, document distribution lists and records, training records, minutes of management reviews or company meetings, and so on)? • Does the organization hold discussions with all levels of employees to communicate the processes of the QMS and its effectiveness?	• It is top management's responsibility to provide the processes and directives so that QMS responsibilities and authorities can be delegated to the appropriate personnel within the organization (where management does not actually determine those roles) and are communicated so that all personnel understand who to contact for approvals and information and to whom to give feedback. • Management must communicate the quality policy, requirements, objectives, and accomplishments to the organization. • Management should actively encourage feedback and communication from people in the organization as a means of involvement. • Management should, in some way, communicate the effectiveness of the QMS to the organization. • It is a new requirement for management to define and implement the communication process.

5.5 Responsibility, Authority and Communication

5.5.3 Internal Communication

Top management must:

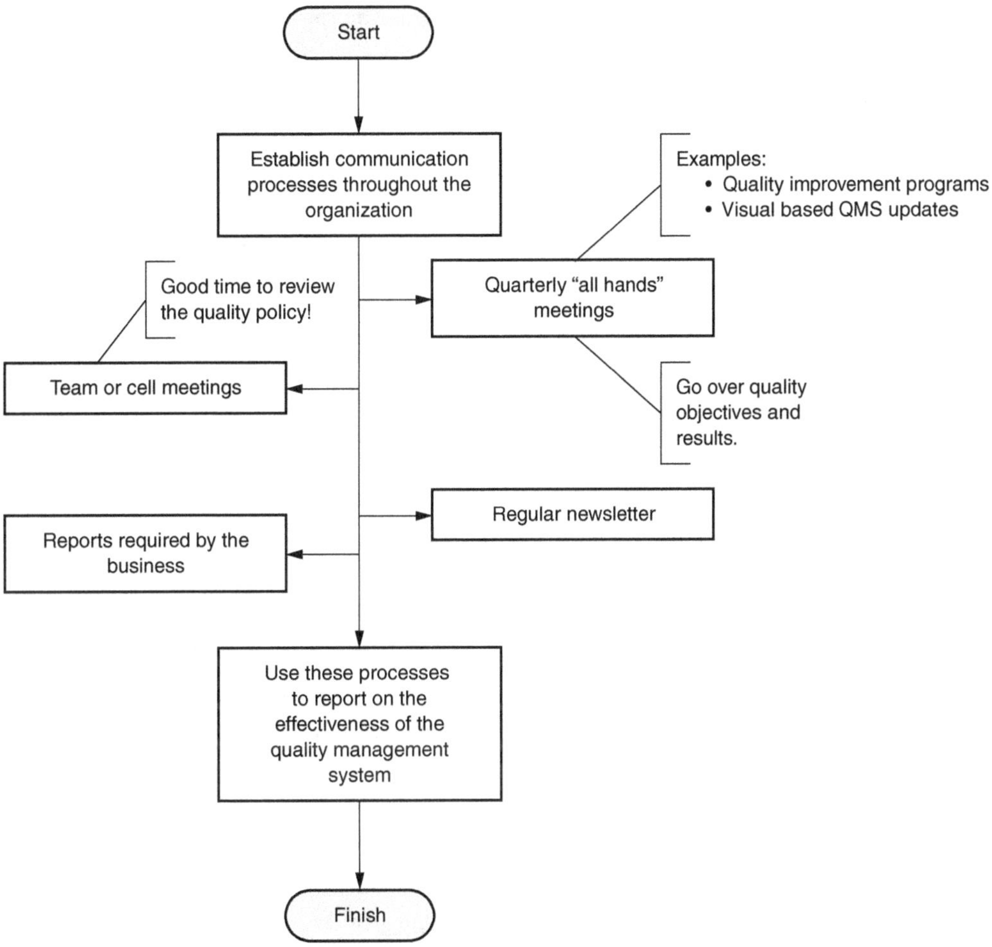

The Standard: 5.0 Management Responsibility

5.6 Management Review

5.6.1 General

13485	Top management is to review the QMS "at planned intervals" to verify that the QMS is appropriate to the organization's needs, satisfies ISO 13485 requirements, and effectively meets customer and regulatory requirements. Meeting minutes of each review must be retained as quality records. During these reviews, management must evaluate potential QMS improvements and changes in response to nonconformities as well as achievements.
9004	Provides guidance on expanding management review to assess QMS efficiency and include review of all activities within the organization. The quality management principles should serve as the basis for "systematic control," with the planning of performance improvements for the whole organization as an end result.
9000	Subclause 2.8.3, Reviewing the quality management system, explores what top management's role is intended to be in the reviews.

Document Requirements:

Requirement:

Records of reviews

Remember:

Management may wish to hold more frequent reviews to address additional items and to demonstrate commitment to continual improvements.

Internal Audit Questions:

- Is there a documented procedure for management reviews?
- Does top management review the QMS at planned intervals to ensure its continuing suitability, adequacy, and effectiveness?
- Do management reviews include evaluation of opportunities for improvement or the need for changes to the organization's quality management system, quality policy, and quality objectives?
- Are results of management reviews recorded and maintained as records?

Management Summary:

- Top management must work with the management representative to establish a periodic review, and must be committed to act in response to the resulting reports.
- Top management should develop an expanded review process that extends to the entire organization.
- Management review should be a platform for exchanging new ideas with open discussion.
- The frequency of management reviews should be determined by the needs of, or challenges to, the organization.
- Management review inputs should be developed to help determine potential changes to the process that will increase the existing efficiency and effectiveness of the QMS.
- Management review minutes are reviewable by regulatory authorities, but all data presented are not. Procedures should be established to identify what must be in the minutes to meet the requirements.

5.6 Management Review

5.6.1 General

Top management must:

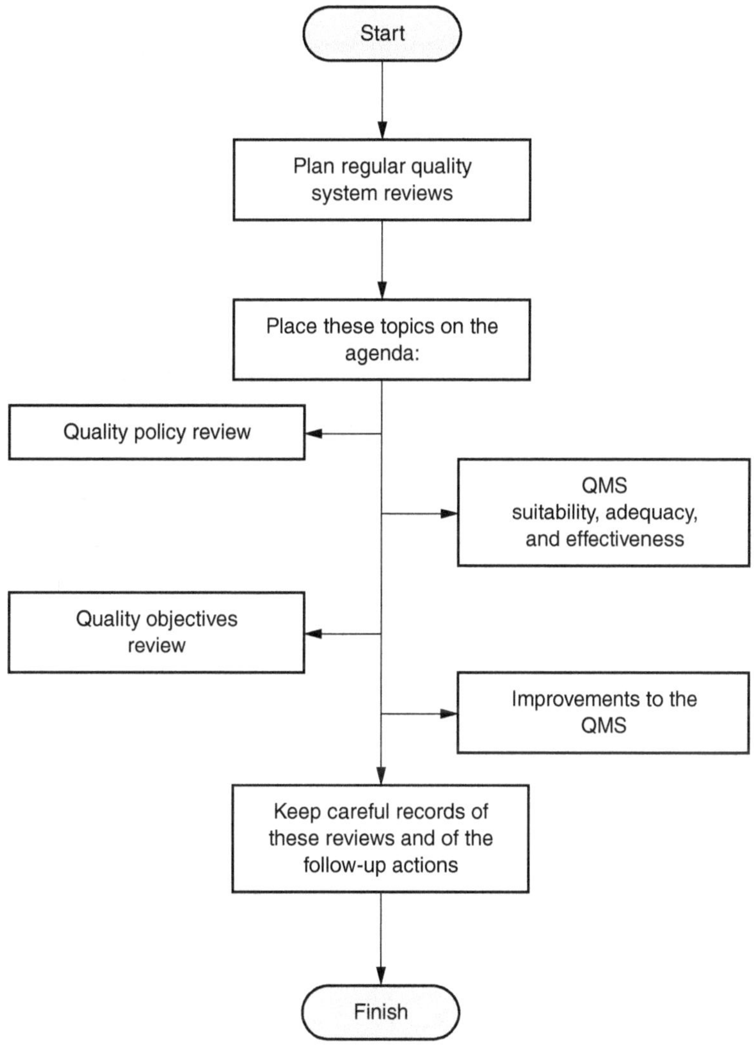

The Standard: 5.0 Management Responsibility

5.6 Management Review

5.6.2 Review Input

13485	Management must include eight categories of data as mandatory review inputs.
9004	Includes as suggested inputs 13 categories of data, ranging from QMS results beyond the ISO 13485 requirements to non-quality business considerations.

Document Requirements:

Requirement:

Records of management review (MR)

Remember:

There are specified MR agenda items (both inputs and outputs) to include within the review. Records of these management reviews should be kept to provide evidence of consideration.

Internal Audit Questions:

- Does management review input include all of the elements in 5.6.2?
- Is top management actively involved within the QMS?

Management Summary:

- Management review inputs should include the status of relevant internal and external factors, as well as results of existing processes.
- Top management should seek verification that QMS effectiveness—and the need for improvements—are known and understood throughout the organization.

5.6 Management Review

5.6.2 Review Input

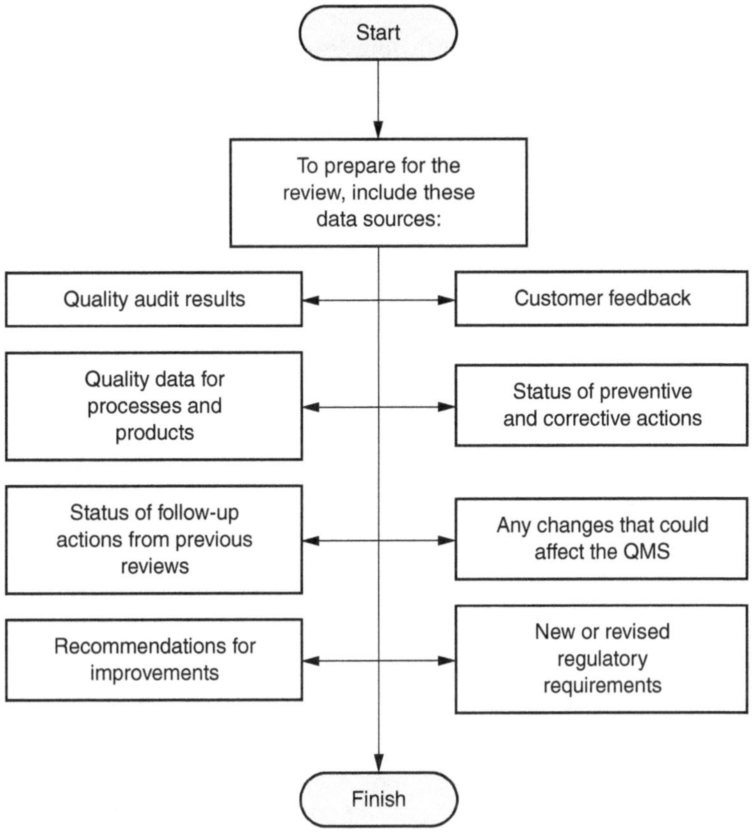

The Standard: 5.0 Management Responsibility

5.6 Management Review

5.6.3 Review Output

13485	The output from management reviews must include any decisions and actions in categories relating to the QMS and customer satisfaction, including product improvements linked to customer requirements.
9004	Provides guidance on how top management can use outputs from an expanded management review to identify potential performance improvements and to motivate employees by establishing new objectives. Suggests six categories of outputs linked to QMS efficiency that could be obtained.

Document Requirements:

Remember:

Management review meeting minutes will be required to document improvement activities and the resources required to initiate these process actions.

Record:

A record of management reviews (including assigned actions) must be documented and retained.

Internal Audit Questions:	*Management Summary:*
• Do management review outputs include actions related to improving the quality management system and its processes? • Do management review outputs include actions to improve product related to customer requirements?	• The strategic planning process, along with the needs of the organization, should determine the frequency of management reviews. • Management reviews should result in changes to the organization's processes so as to increase the existing efficiency and effectiveness of the QMS. • Adequate records should be kept to provide traceability and facilitate the evaluation process in future management reviews.

5.6 Management Review

5.6.3 Review Output

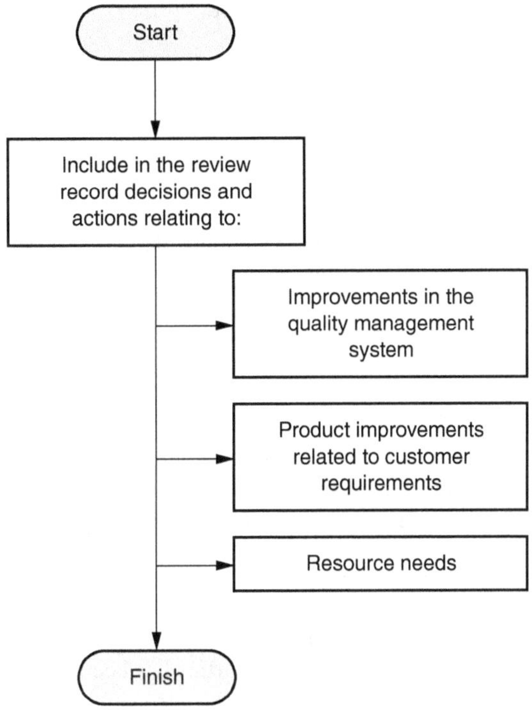

Clause	Change	Interpretation
5.1 (Management commitment)	*Note* present in the 2003 version, relating to statutory requirements for safety and performance of the medical device, has been removed.	*Note* is no longer relevant, as regulatory requirements in addition to the assessment of risk are now included throughout the 2016 standard.
5.2 (Customer focus)	Addition of applicable regulatory requirements.	Ensures that applicable regulatory requirements are applied to address a client's approach to customer focus.
5.4.1 (Quality objectives)	Addition of requirement that quality objectives must meet applicable regulatory requirements.	Organizations must now ensure that regulatory requirements are met and documented as part of the client's quality objectives.
5.5.1 (Responsibility and authority)	*Note* relating to national or regional regulations requiring a person to be assigned to monitor post-production activities, for example, post-market surveillance and adverse events, has been removed from the 2016 version of the standard.	Removed *Note* is now covered by the responsibilities of the authorized representative, where applicable.
5.5.2 (Management representative)	*Note* relating to the responsibility of the management representative liaising with external parties has been removed.	This *Note* has been removed since this is now covered in the 2016 version, particularly in 8.2.3, Reporting to regulatory authorities.
5.6 (Management review)	Management review process must be documented (procedure) and must be held at defined intervals.	Organizations must now have a documented procedure for management review, which must also cover the frequency at which the meetings are held.
5.6.3 (Review output)	Management review output is now required to cover changes needed to respond to applicable new or revised regulatory requirements.	Organizations must now document any changes required in order to respond to applicable new or revised regulatory requirements, for example, impact to client QMS review, or management of necessary resources.

SUMMARY

by S. Pletzer

ISO 13485:2016—5 Management Responsibility

5.1 Management commitment

Top management provides evidence of their commitment to developing and implementing an effective QMS.

5.2 Customer focus

Top management determines the needs of the QMS in order to meet customer requirements.

5.3 Quality policy

Top management establishes an appropriate and effective quality policy, which is communicated to the organization and reviewed for suitability.

5.4 Planning

Top management ensures that quality objectives are measurable and consistent with quality policy, and changes to the system are planned and maintained.

5.5 Responsibility, authority and communication

Top management defines and communicates responsibilities and authorities, establishes personnel interrelations, and appoints a management member as representative.

5.6 Management review

Top management reviews the organization's QMS at planned intervals, reviews various inputs, makes decisions based on the review of these inputs, and maintains records of inputs and outputs at these intervals.

SECTION 6: RESOURCE MANAGEMENT

The Standard: 6.0 Resource Management

6.1 Provision of Resources

13485	The organization must identify and make available the resources for the organization and ongoing increases in the QMS, and improvement in the level of customer satisfaction and regulatory compliance.
9004	Defines possible resources to be provided and recommends that top management provide the resources to pursue strategic goals and achieve objectives. Includes a list of possible resources and considerations to help improve organizational performance.

Document Requirements:

Remember:

Management will be required to formalize its resource planning process to ensure timely availability.

Internal Audit Questions:

- Has the organization determined and provided the resources necessary to implement the processes of the quality management system?
- Has the organization determined and provided the resources necessary to improve the processes of the quality management system?
- Has the organization determined and provided the resources necessary to address customer satisfaction and regulatory requirements?
- Are the resources provided in a timely manner?
- What sources are reviewed for potential resource needs?

Management Summary:

- Top management should provide adequate resources to implement the QMS and achieve its objectives.
- Top management must direct those having responsibility and/or authority for the QMS and related organizational operations to provide specified inputs to be used in management reviews of the QMS, including the frequency of such inputs and the processes and forms in which the inputs are to be provided.
- Top management should consider relevant inputs (for example, management review agenda items) to be reviewed for resource needs.
- Top management should approach evaluation of resources with consideration of what the quality policy statement and quality policy objectives state, and what the organization's business performance measurements indicate (that is, the KPIs).
- Resources may include:
  - People
  - Infrastructure
  - Work environment
  - Information
  - Suppliers and partners
  - Natural and financial resources

6.1 Provision of Resources

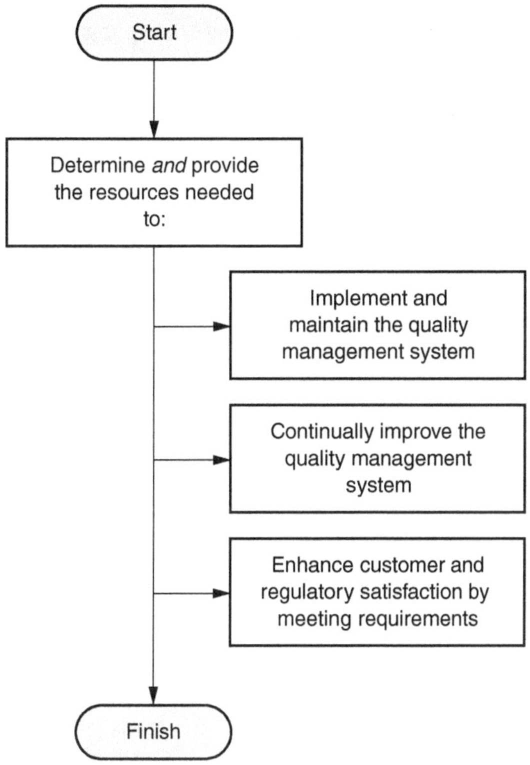

The Standard: 6.0 Resource Management

6.2 Human Resources

13485	Personnel whose actions impact product quality must be competent to work on products, based on their education, training, skills, and experience.
9004	For an organization to achieve performance improvements, ISO 9004 recommends 12 types of activities to pursue to encourage the involvement and development of its people.

Document Requirements:

Remember:

- Records that are created by the activities to assure competency may need to be controlled per subclause 4.2.5, Control of records. *See also* 4.2.4, Control of documents
- Records of training. While the need for documentation of any decisions reached and actions determined as outputs of the reviews is not specified, some form of documentation will be necessary in most cases to communicate the improvements that top management wants to the QMS and/or product.

Internal Audit Questions:

- Do personnel with assigned responsibilities within the quality management system and organization have competency in the appropriate education, training, skills, and experience?
- Has top management made changes to the quality policy and/or quality objectives related to personnel as a result of management reviews?

Management Summary:

- Improvements to the effectiveness and efficiency of the organization should be accomplished through the selection, involvement, and support of the people within the organization.
- Top management must use management reviews as a forum in which to ensure that the QMS remains effective in meeting the needs of the organization and its customers, and to determine what continual improvement efforts will be undertaken before the next review.
- It is expected that top management will reach decisions about improvements and direct the organization to take steps to achieve those improvements as outputs of management reviews.

6.2 Human Resources

Anyone who can affect product quality must be competent to do so.

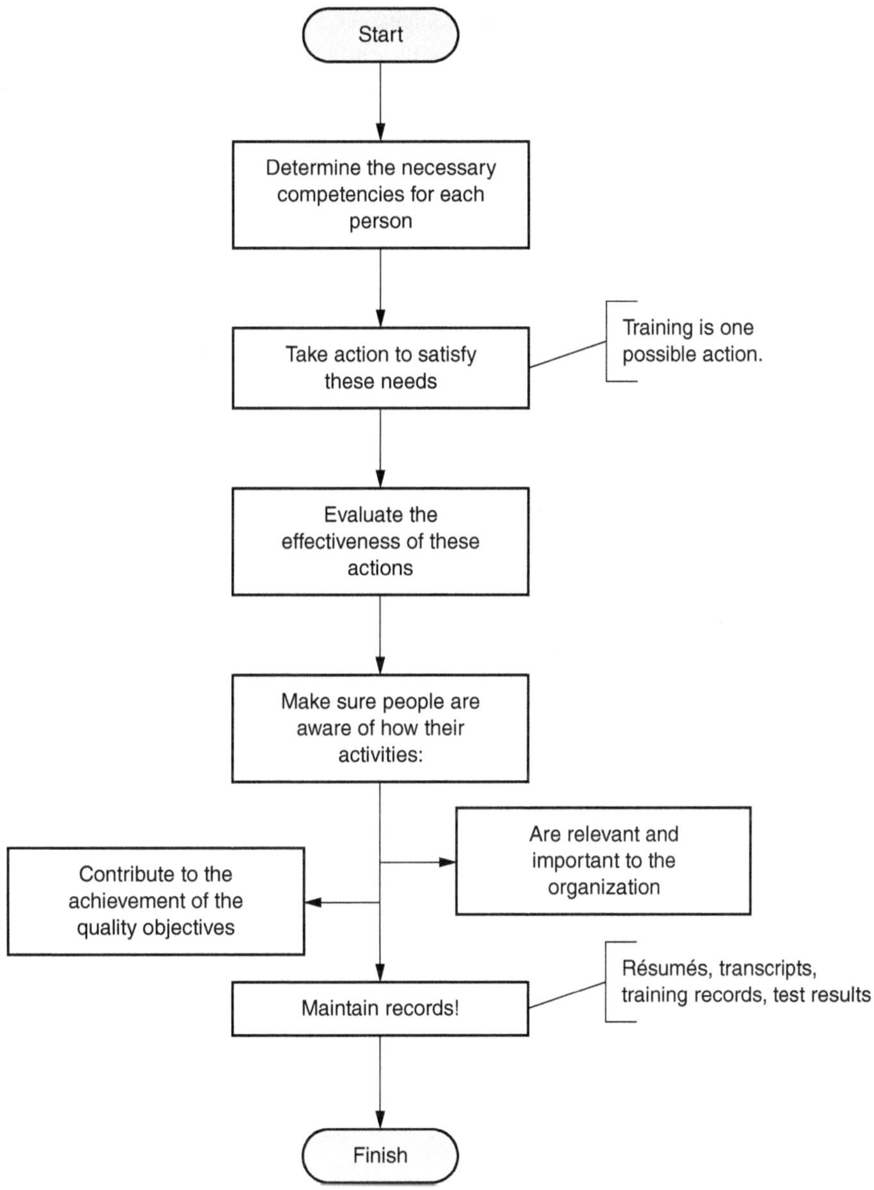

The Standard: 6.0 Resource Management

6.2 Human Resources

Continued . . .

13485 The organization must identify the qualifications required for each position affecting product quality, take appropriate actions to ensure personnel competency in those positions, assess the effectiveness of these efforts, and make personnel aware of how they affect quality and what they can do to achieve quality objectives. The organization must document personnel competence and actions taken relative to competence and retain these documents as quality records.

9004 Recommends that an organization analyze present and future needs and determine gaps between needs and existing personnel capabilities, with consideration of five sources for determining competency needs. Explores what to consider in terms of education and training relative to meeting quality objectives and raising awareness. Defines possible subjects to include in education and training and to address in training plans.

Document Requirements:

Required:

Record (e)

Internal Audit Questions:

- Does the organization identify the competency needs of personnel performing activities affecting quality (including specialized training needs such as certifications)?
- Does the organization provide training to satisfy these needs?
- Does the organization evaluate the effectiveness of the training?
- How is this done?
- Does the organization ensure that employees are aware of the importance of their activities and how they contribute to the achievement of the quality objectives?
- How is this done?
- Does the organization have provisions for training of temporary employees?
- Does the organization maintain records of education, experience, training, and qualifications (that is, banners, posters, bulletin boards, or intranet sites)? Does the organization maintain similar records for temporary employees?

Management Summary:

- Management should analyze both the present and expected competence needs within the organization.
- Management should plan for education and training needs and consider the organization's processes, stages of development, and culture.

6.2 Human Resources

Anyone who can affect product quality must be competent to do so.

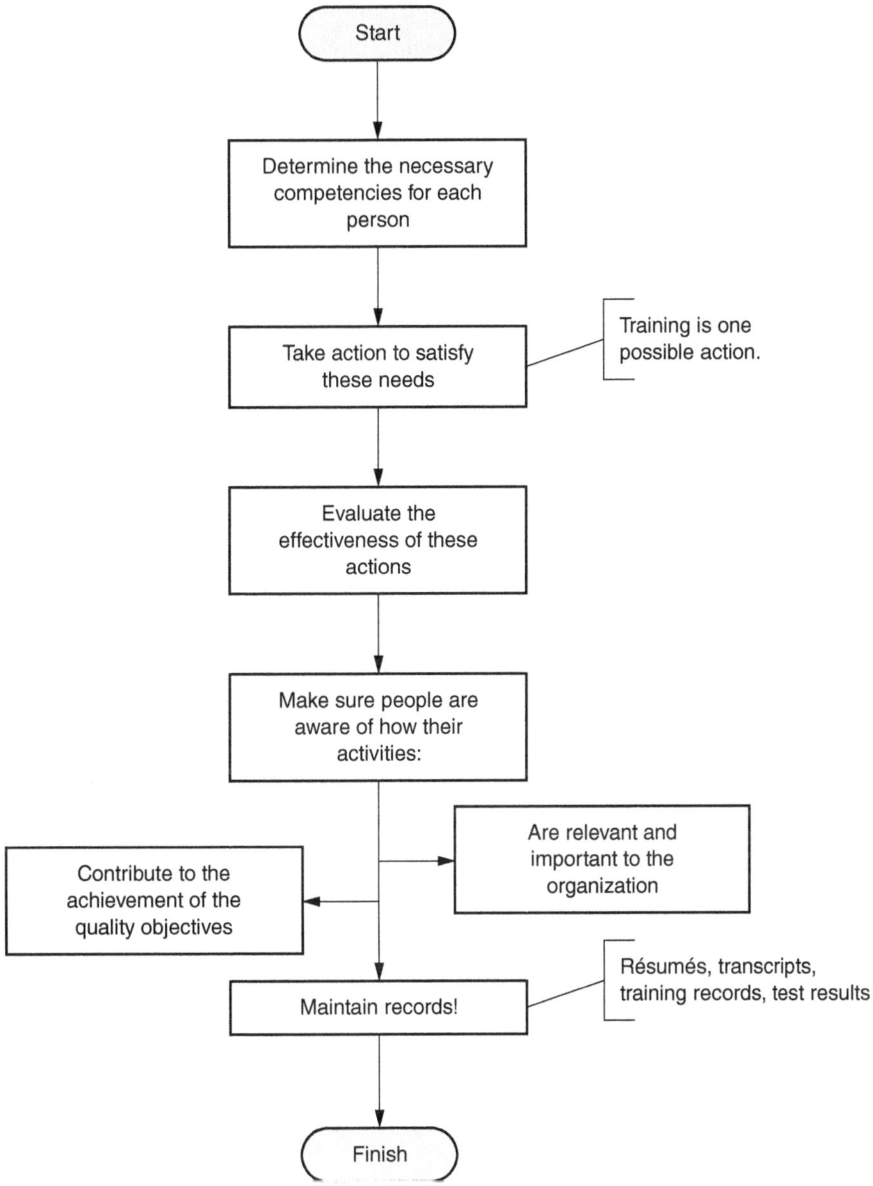

The Standard: 6.0 Resource Management

6.3 Infrastructure

13485	The organization is responsible for the establishment and upkeep of the infrastructure required to provide customers with products according to specifications, with examples of what is meant by *infrastructure* spelled out.
9004	Recommends four factors that management should address in defining the required infrastructure, and advises an organization to consider natural phenomena and the risks they pose to its operations in developing and keeping up the infrastructure.
9000	Defines *infrastructure* in 3.3.3 as a "system of facilities, equipment and services needed for the operation of an organization."

Document Requirements:

Remember:

The records that are created by the activities performed to fulfill the requirements of this clause may need to be controlled per subclause 4.2.5, Control of records. *See also* 4.2.4, Control of documents.

Internal Audit Questions:

- Does the organization identify, provide, and maintain the work space and associated facilities needed to achieve conformity of product?
- Where and how is this done?
- Does the organization identify, provide, and maintain the supporting services, equipment, hardware, and software it needs to achieve conformity of product?
- For key areas that can impact quality (such as calibration, preventive maintenance, pest control, and so on), are there written procedures that address the requirements of such maintenance activities and their frequency?

Management Summary:

- Infrastructure includes plant, utilities, work space, tools, equipment, support services (such as pest control, and so on), information/communication technology, and transport facilities.
- Equipment activities include, at a minimum, validation planning and execution, maintenance, calibration, recovery, and so on.

6.3 Infrastructure

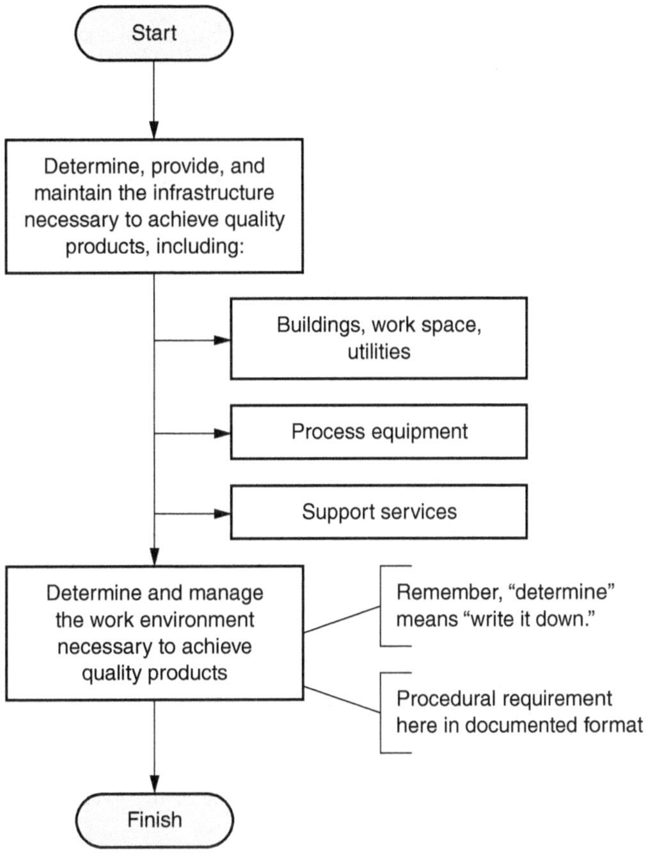

The Standard: 6.0 Resource Management

6.4 Work Environment and Contamination Control

13485	The organization is required to maintain the work environment in a state that permits product to satisfy requirements. The organization is required to have procedures for personnel, even temporary personnel, and the work environment to preserve the quality of the product and, when appropriate, the isolation of potentially contaminated product.
9004	Encourages management to maintain the work environment so as to improve employee motivation, satisfaction, and performance, and thereby improve organizational performance. Identifies seven aspects of the work environment that should be evaluated in developing or revising the work environment.
9000	Defines *work environment* in 3.3.4 as a "set of conditions under which work is performed," with a list of conditions to be included in the scope of the work environment. Clarifies this as the conditions under which work is performed, not conditions under which people perform work.

Document Requirements:

Remember:

The records that are created by the activities performed to fulfill the requirements of this clause may need to be controlled per subclause 4.2.5, Control of records. *See also* 4.2.4, Control of documents.

Internal Audit Questions:

- Does the organization identify and manage the human and physical factors of the work environment needed to achieve conformity of product?
- Does the organization have procedures (personal hygiene, housekeeping, and so on) to ensure the preservation of product quality?
- Does the organization have procedures for monitoring controlled areas?
- Does the organization require that temporary employees working under special environmental conditions are appropriately supervised and/or trained?
- Does the organization have a procedure for control of product potentially impacted by being processed in an unsuitable environment?
- Does the organization have a procedure for control of potentially contaminated product?

Management Summary:

- Using a combination of human and physical factors, management should consider:
  - Creative work methods and opportunities for greater involvement
  - Ergonomics
  - Heat, humidity, light, and airflow
  - Hygiene, cleanliness, noise, vibration, and pollution
  - Product flow through the manufacturing process
  - Impact to product processed in unsuitable environments
  - Utilization strategies for temporary employees (including visitors) in "special environment" areas
- For sterile medical devices, the organization shall document requirements for control of contamination with microorganisms or particulate matter, and maintain the required cleanliness during assembly or packaging processes.

6.4 Work Environment and Contamination Control

6.4.1 Work Environment

6.4.2 Contamination Control

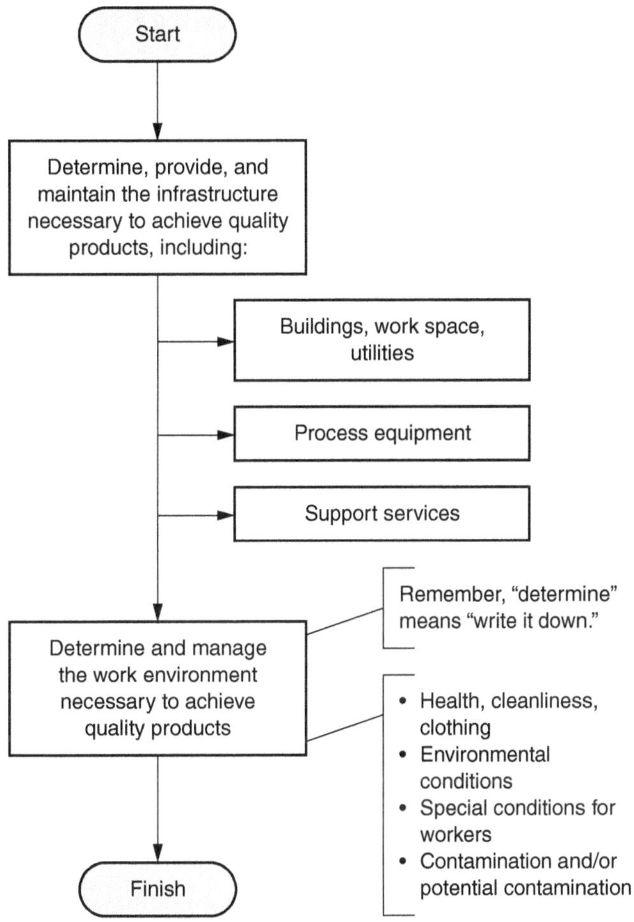

Clause	Change	Interpretation
6.2 (Human resources)	Additional requirement to document the processes for establishing competence, providing training, and ensuring awareness. There is also a link to the method used to measure effectiveness based on the risk of the work or role for which the training has been provided.	Organizations must document how they assess competency for the roles in their organization and also show how the training process is performed, for example, by new hires or current employees. The process for measuring competency, for example, with a skills matrix, must also be based on the risk that the employees' activities pose to the product.
6.3 (Infrastructure)	Additional requirement to ensure that infrastructure is managed to prevent product mix-up. In addition to the maintenance of equipment being applied to production, this means control of the work environment and monitoring and measurement equipment.	Organizations must have a documented procedure that details how the infrastructure is managed in order to prevent product mix-up, for example, detailing specific areas for performing key activities such as final inspection and quarantine. The procedure must document how these areas are managed (building management systems) to ensure that product conformity is maintained.
6.4.1 (Work environment)	Additional requirements for documentation of the work environment conditions, competency of temporary work performed, and a *Note* referring to ISO 14644 and ISO 14698 for further information.	Following on from clause 6.3, these are additional documentation requirements that must be adhered to for managing the work environment, with a specific *Note* for clients to refer to ISO 14644—*Cleanrooms and associated controlled environments* and ISO 14698—*Cleanrooms and associated controlled environments—Biocontamination control.*
6.4.2 (Contamination control)	New requirement to control contaminated or potentially contaminated product to prevent contamination of the work environment, personnel, or product.	Organizations must control contaminated product, for example, returned complaint samples or contaminated material from supplier. Clients must have documented procedures in place to prevent contamination from spreading throughout product realization processes.

SUMMARY

by S. Pletzer

ISO 13485:2016—6 Resource Management

6.1 Provision of resources

The organization provides resources to implement and maintain the QMS and to meet regulatory and customer requirements.

6.2 Human resources

The organization determines personnel competency, ensures awareness, and maintains training records.

6.3 Infrastructure

The organization provides and maintains the infrastructure to meet product requirements, including maintenance activities.

6.4 Work environment and contamination control

6.4.1 Work environment

6.4.2 Contamination control

The organization determines and manages the work environment to meet product requirements. For sterile medical devices, the organization shall document requirements for control of contamination with microorganisms or particulate matter, and maintain the required cleanliness during assembly or packaging processes.

SECTION 7: PRODUCT REALIZATION

- 7.1 Planning of product realization
- 7.2 Customer-related processes
  - 7.2.1 Determination of requirements related to product
  - 7.2.2 Review of requirements related to product
  - 7.2.3 Communication
- 7.3 Design and development
  - 7.3.1 General
  - 7.3.2–10 Design and development
- 7.4 Purchasing
  - 7.4.1 Purchasing process
  - 7.4.2 Purchasing information
  - 7.4.3 Verification of purchased product
- 7.5 Production and service provision
  - 7.5.1 Control of production and service provision
  - 7.5.2 Cleanliness of product
  - 7.5.3 Installation activities
  - 7.5.4 Servicing activities
  - 7.5.5 Particular requirements for sterile medical devices
  - 7.5.6 Validation of processes for production and service provision
  - 7.5.7 Particular requirements for validation of processes for sterilization and sterile barrier systems
  - 7.5.8 Identification
  - 7.5.9 Traceability
  - 7.5.10 Customer property
  - 7.5.11 Preservation of product
- 7.6 Control of monitoring and measuring equipment

The Standard: 7.0 Product Realization

7.1 Planning of Product Realization

13485 The organization is to engage in planning to create processes to achieve product realization—the processes required to design, develop, produce, deliver, and/or service a product—in conformance with other QMS requirements. Outcomes are to be identified as a result of the planning process, including a product's quality objectives; processes, documentation, and resources for the product; risk management procedures addressing each stage of product realization; procedures that will ensure the product meets specifications; records that prove the processes and product satisfy requirements. The outputs of the planning process must be appropriate for use by the organization, based on how it functions.

9004 Recommends that top management take steps to ensure the effectiveness and efficiency of all product realization and support processes, and examine these processes as both inputs and outputs, with consideration of the roles of documentation, personnel, and operating plans. Examples are provided of support processes, input issues, and topics to be covered in process performance reviews.

9000 Defines *inspection*, *test*, *verification*, and *validation* in 3.8.2–3.8.5.

Document Requirements:

Required:

Record (d)

Remember:

A documented procedure may be needed for describing how the planning is accomplished.

Internal Audit Questions:

- Have documented quality plans for product realization processes and product validation been established (that is, risk management plan, drawings, specifications, materials, process flow diagrams, process flowcharts, setup sheets, validation reports, and so on)?
- Is there evidence for the planning of production processes?
- Does the planning encompass all product realization processes?
- Is the planning consistent with other elements of the quality management system?
- Does documentation of product realization exist?
- Are product realization resources and facilities defined during the planning process? Are they adequate?
- Does planning define the records that are kept to show confidence in the conformity of the processes and final product?

Management Summary:

- For effective operation, management should recognize that a process output might become the input to another process.
- Verification results and process validations should be considered as inputs to a process.
- Effective processes offer increased benefits, improved customer satisfaction and use of resources, and waste reduction.
- Risk identification and management must be included in planning for product realization.
- Process inputs can be internal or external to the organization, such as:
  - Risks and product requirements
  - Competence of people
  - Documentation
  - Equipment capability and monitoring
  - Health, safety, and work environment

7.1 Planning of Product Realization

The organization must define its own processes for product realization.

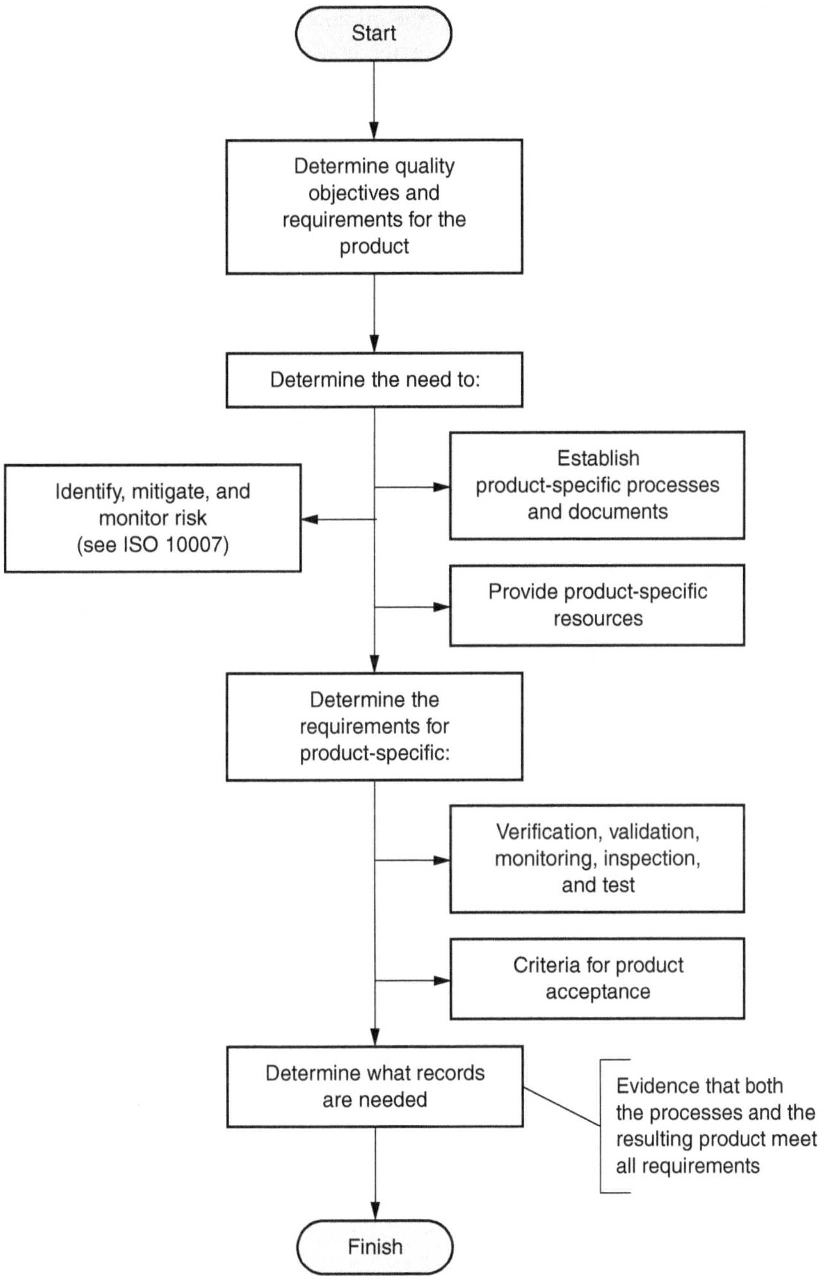

The Standard 7.0 Product Realization:

7.2 Customer-Related Processes

7.2.1 Determination of Requirements Related to Product

13485	The organization must identify all requirements of a product, including those from customer specifications, unspecified but necessary for the product to work properly, imposed by legal and regulatory authorities, or deemed necessary by the organization.
9004	Recommends that management consider the needs and expectations of all interested parties, including the customer, in determining other requirements for a product. (Note: This also applies for 7.2.2 and 7.2.3.)
9000	Defines *requirement* in 3.1.2 as a "need or expectation that is stated, generally implied or obligatory," with four *Summaries* clarifying the definition.

Document Requirements:

Remember:

The organization should consider having a documented procedure for determining customer requirements. A documented procedure is a vital part in achieving customer requirements, needs, and expectations.

Internal Audit Questions:

- Has the organization determined customer requirements?
- Have processes been established to determine both specified and unspecified customer requirements?
- Has the organization determined applicable regulatory or other requirements? How is risk determined and documented for the product?
- Do records exist to provide evidence that customer requirements have been determined (that is, contract review records, service records, customer feedback records, surveys, market testing reports, design input and design validation records, identified legal and regulatory requirements, and so on)?

Management Summary:

Management should consider the following to ensure that the customer's needs and expectations are met:

- Operating conditions for the product
- Use or application of the product
- Risk associated with the product
- Disposal of the product
- Life cycle of the product
- Environmental impact of the product

7.2 Customer-Related Processes

7.2.1 Determination of Requirements Related to Product

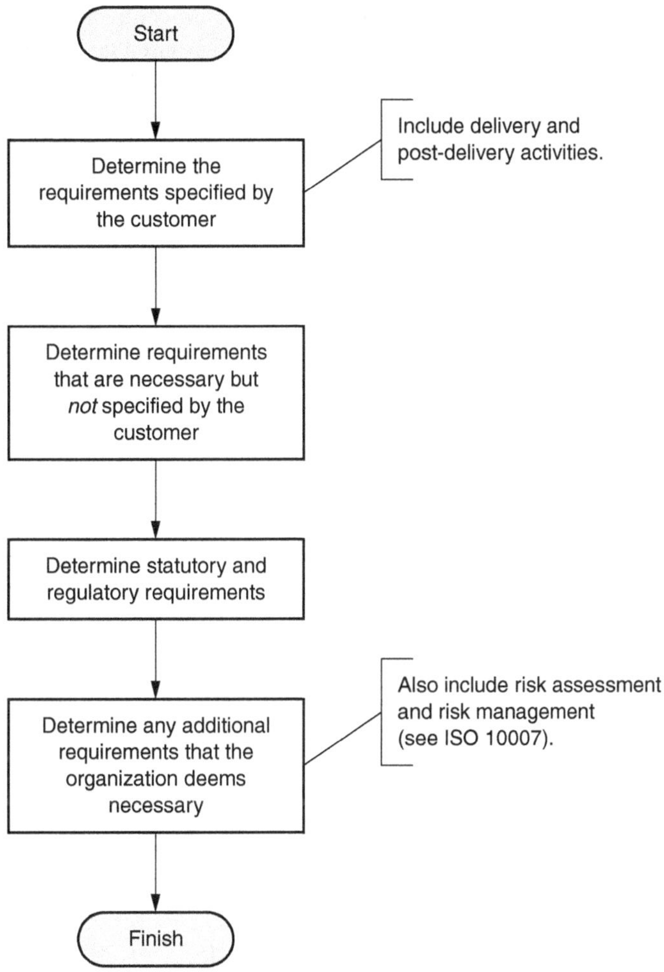

The Standard: 7.0 Product Realization

7.2 Customer-Related Processes

7.2.2 Review of Requirements Related to Product

13485 Product requirements, whether involving a bid request, a new contract order, or an amendment to an existing contract, must be reviewed by the organization to ensure that the requirements and/or any changes are clearly understood and can be met, including in situations where the customer provides no written specifications. How an organization does this must be defined and documented. The review process results are to be treated as quality records, which must be updated whenever requirements change—with changes relayed to relevant personnel to ensure that customer requirements are met effectively.

9000 Defines *review* in 3.8.7 as "activity undertaken to determine the suitability, adequacy and effectiveness of the subject matter to achieve established objectives."

Document Requirements:

Required: N/A

Record

Remember:

Consider having a documented procedure for the review of customer requirements and how these requirements can be met.

Internal Audit Questions:

- Is there a procedure that defines the requirements for the review of identified customer requirements before commitment to supply a product to the customer?
- How are these requirements documented?
- Is there a procedure that requires the review of quotes and orders for adequacy of the definition of requirements?
- Is there a process for handling review of verbal orders?
- Is there a process to handle resolution of differences between quotations and orders?
- Is there a process for handling changes to product requirements?
- Are records of quote, tender, contract, and order review available and maintained?

Management Summary:

- Product and process changes should be identified, recorded, evaluated, reviewed, and controlled in order to understand their effect on other processes and the needs and expectations of customers. The process for doing this should be documented.
- The organization should define the authority for initiating process changes in order to maintain control.

7.2 Customer-Related Processes

7.2.2 Review of Requirements Related to Product

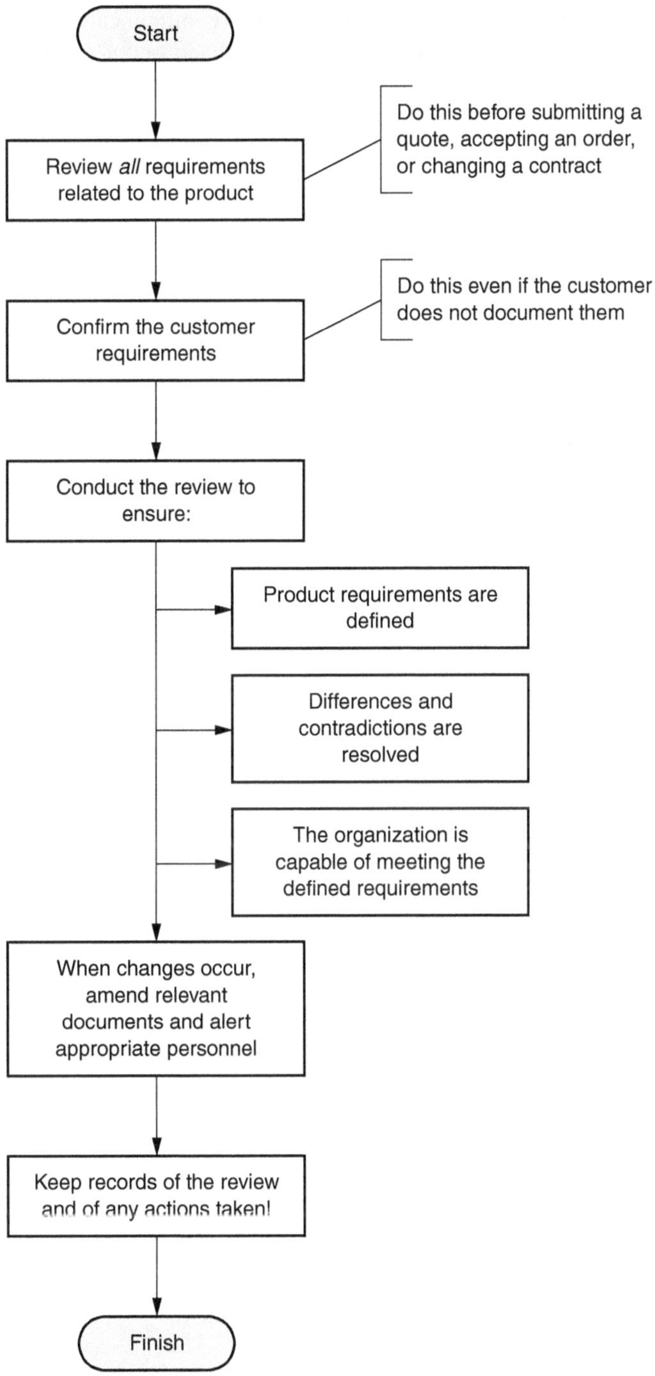

The Standard: 7.0 Product Realization

7.2 Customer-Related Processes

7.2.3 Communication

13485 The organization must establish procedures for effective communication with customers regarding product specifications, advisory notices, the receipt and handling of product orders, including amendments, and any questions regarding these, and in obtaining customer feedback on products and related services.

Document Requirements:

Remember:

Feedback records should be evident.

Internal Audit Questions:	*Management Summary:*
Is there a process in place to communicate with customers regarding product information (design inputs), advisory notices, inquiries, contracts, order handling (including amendments), and customer feedback, including customer complaints?	• The organization should implement and maintain a process to translate customer needs and expectations into requirements for the organization. • The organization should have established means of communicating information to customers. • The organization should fully understand the customer requirements prior to acceptance of orders. • The organization should have established means of identifying, receiving, and, where appropriate, documenting communications from customers.

7.2 Customer-Related Processes

7.2.3 Communication

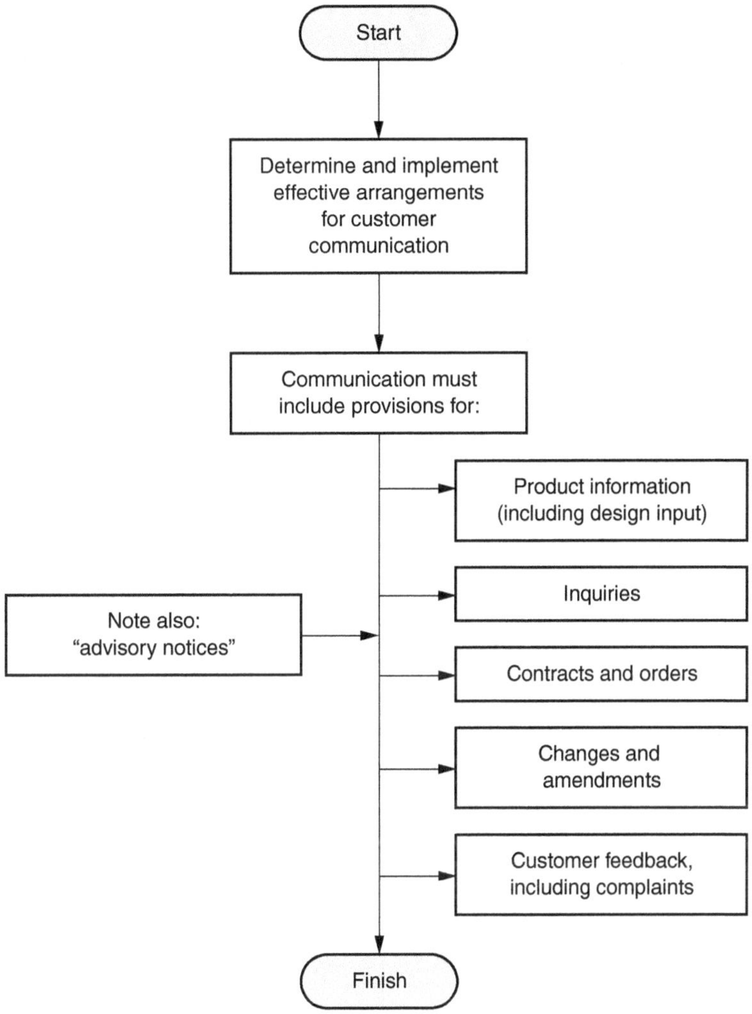

The Standard: 7.0 Product Realization

7.3 Design and Development

7.3.1 General

7.3.2 Design and Development Planning

13485	The processes involved in design and development of a product must be planned out, managed, and documented by the organization to ensure that each group involved with its design and development is engaged at all appropriate stages—and everyone involved understands who is responsible for which activities at what times—and that proper evaluations of prototypes, and so on, are conducted at appropriate stages. The output from this planning must be revised to reflect the product's progress through the design and development process.
9004	Suggests that top management should consider a number of aspects and techniques not directly related to customer specifications for the product in planning its design and development, including the design of the organization's processes. Recommends taking steps to identify and mitigate potential risk to product users and those affected by the organization's processes, providing a list of six risk assessment tools that could be applied in planning design and development.
9000	Defines *design and development* in 3.4.4 as a "set of processes that transforms requirements into specified characteristics or into the specification of a product, process or system."

Document Requirements:

Required:

A procedure and multiple records throughout 7.3 should be evident.

Internal Audit Questions:

- Does the organization have procedures for the design and development of processes, products, or services?
- Are the design and/or development project stages defined? Where?
- Are verification, validation, and design transfer activities addressed and appropriate?
- Is it clear who is responsible for what?
- Are design and development inputs and outputs defined?
- Are the communication channels defined?
- Does evidence exist to show that communication on projects is occurring and is effective?

Management Summary:

- Risk assessment should be utilized to assess the potential for, and the effect of, possible failures or faults in products or processes.
- Management should consider the following factors that contribute to meeting the product and process performance expected by customers:
  - Life cycle
  - Safety and health (including sterility)
  - Testability and usability
  - User-friendliness
  - Dependability and durability
  - Delivery
  - Process capability
  - Ergonomics
  - Communication
  - Product disposal and the environment

7.3 Design and Development

7.3.1 General

7.3.2 Design and Development Planning

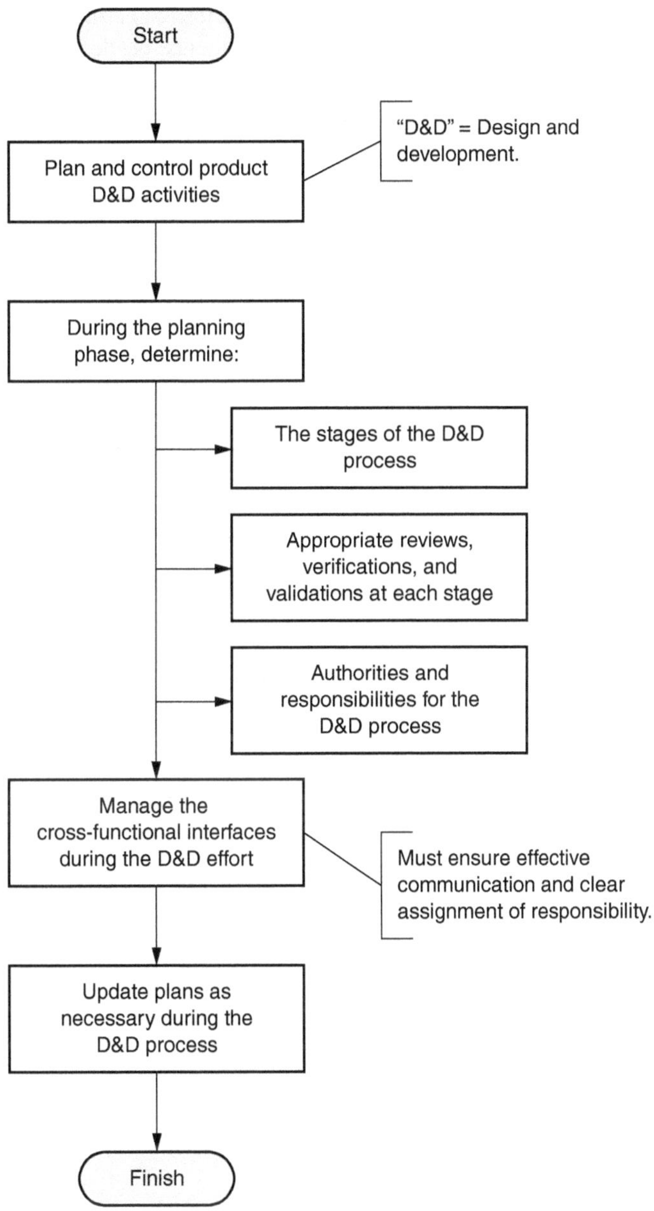

The Standard: 7.0 Product Realization

7.3 Design and Development

7.3.3 Design and Development Inputs

13485 The organization is to identify all inputs required for the design and development process, evaluate their adequacy for determining product design and development, and treat the identified inputs for a product as a quality record. Four specific types of inputs are required to be included (five for medical devices—[a]–[e]), and the inputs identified must cover all requirements, be clearly defined, and cannot create conflicting situations.

9004 Recommends identification of process as well as product inputs and coupling of internal needs and expectations with external ones. Presents examples of external inputs, internal inputs, and inputs that identify critical process and product characteristics.

Document Requirements:

Record required

Internal Audit Questions:

- Are all product requirements defined and documented?
- Are the requirements complete, unambiguous, and without conflict?
- How are these requirements documented, reviewed, and approved?

Management Summary:

- External and internal needs and expectations should be evaluated and deemed suitable for translation into input requirements for design and development processes.
- For medical devices, product functionality, performance and safety requirements, as well as applicable regulatory requirements and risk management outputs for the intended use must all be considered.

7.3 Design and Development

7.3.3 Design and Development Inputs

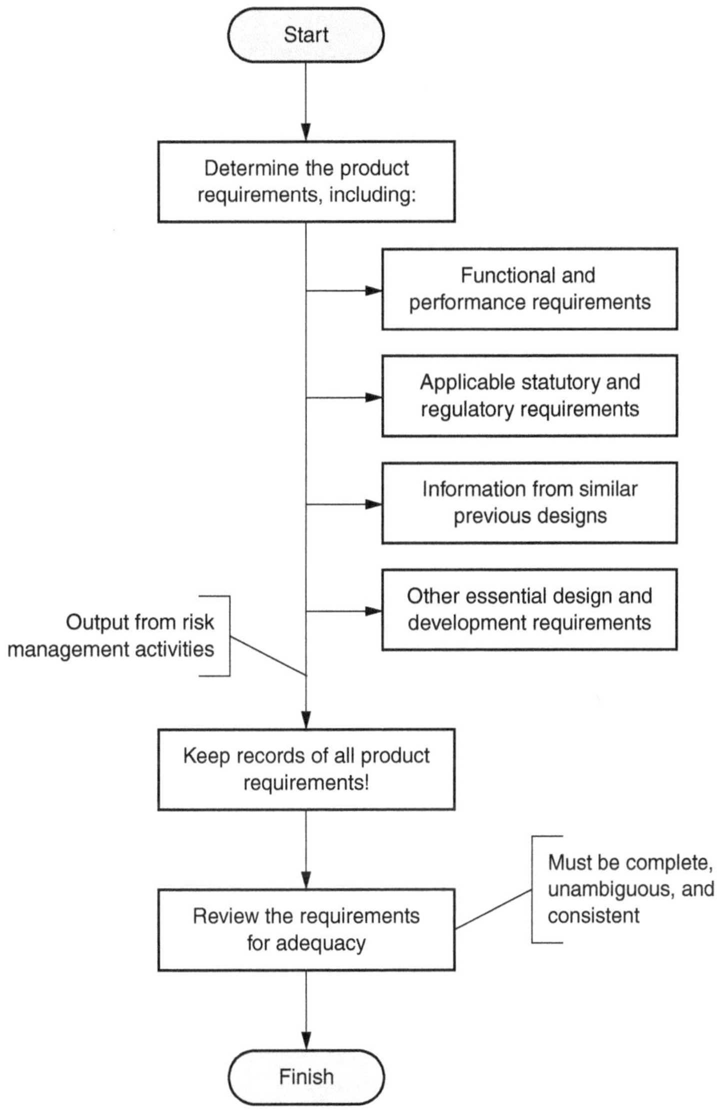

The Standard: 7.0 Product Realization

7.3 Design and Development

7.3.4 Design and Development Outputs

13485	The organization must capture and approve the outputs in a medium that permits verification against the inputs before their release. Design and development outputs must meet four criteria ([a]–[d]), including the ability to satisfy requirements identified as inputs, disseminate data to accomplish product-related processes within the organization, and spell out product characteristics that are critical to its safe and proper use.
9004	Provides examples of eight types of outputs.

Document Requirements:

Record

Internal Audit Questions:	*Management Summary:*
• Does the output of design and/or development projects exist? • Does it show how design and/or development outputs satisfy input requirements? • Does the output provide information for production operations? • Are product acceptance criteria clearly stated? • Are product safety and use characteristics identified? • Does a documentation and approval process exist for the release of products from the design and/or development process?	• Design and development outputs should be reviewed against inputs to provide objective evidence that they have effectively met the process and product requirements. • The design outputs and their review should be documented.

7.3 Design and Development

7.3.4 Design and Development Outputs

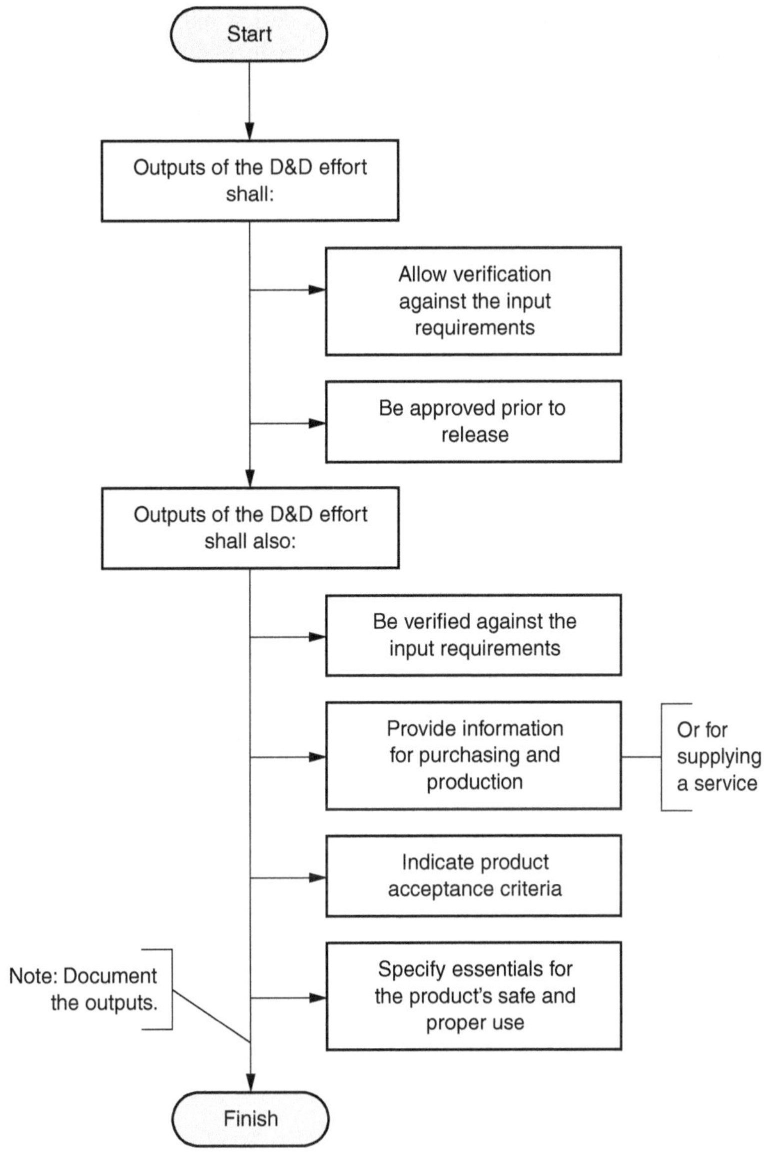

The Standard: 7.0 Product Realization

7.3 Design and Development

7.3.5 Design and Development Review

13485	The organization is to engage in systematic reviews of design and development activities at appropriate stages ([a]–[b]) to ensure that the results are meeting product and design and development requirements and to catch and correct problems before they impact overall design and development. All groups involved in design and development should be represented in these reviews, and the review results are to be treated as quality records.
9004	Provides examples of topics to be covered during the reviews.

Document Requirements:

Record

Internal Audit Questions:	*Management Summary:*
• Are reviews of design and/or development being performed? Are they indicated in the project planning documents? • Is there a systematic approach to when these reviews are performed? • Who attends these reviews? Is the attendance appropriate, and are roles defined and documented? • Are results of the reviews documented? Are follow-up actions taken?	• Systematic reviews may be conducted at suitable stages in the design and development process, as well as at completion. • Participants and their defined roles (including that of an independent reviewer—from 21 CFR 820) should be commensurate with the stage or items being reviewed.

7.3 Design and Development

7.3.5 Design and Development Review

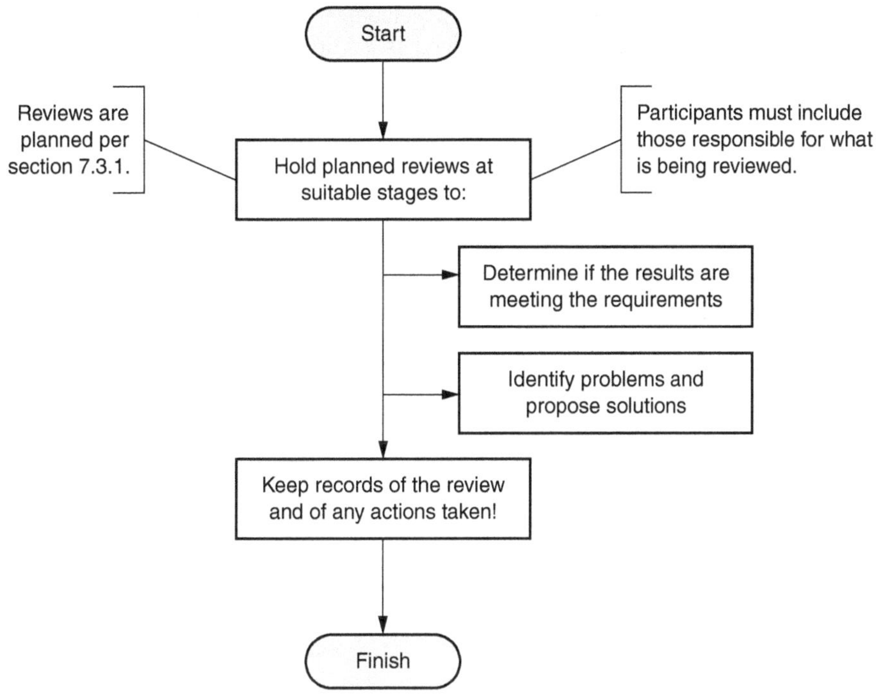

The Standard: 7.0 Product Realization

7.3 Design and Development

7.3.6 Design and Development Verification

13485	The organization must verify that design and development outputs meet all input requirements, and that the results of verification activity are treated as quality records.
9004	Provides examples of verification and validation activities to be undertaken to ensure that the end product will be well received by customers and others.
9000	Defines *verification* in 3.8.4 as the "confirmation, through the provision of objective evidence, that specified requirements have been fulfilled," with activities that constitute confirmation provided in a *Note*.

Document Requirements:

Record (in addition, a procedure, while not required, is encouraged).

Internal Audit Questions:	*Management Summary:*
• Is there a verification process in place? • Is it effectively implemented? • Are results of the verification documented? • Are follow-up actions recorded?	The organization should evaluate design and development outputs and processes in order to prevent nonconformities and deficiencies. This may include: • Design verification • Process validation • Sterilization validation

7.3 Design and Development

7.3.6 Design and Development Verification

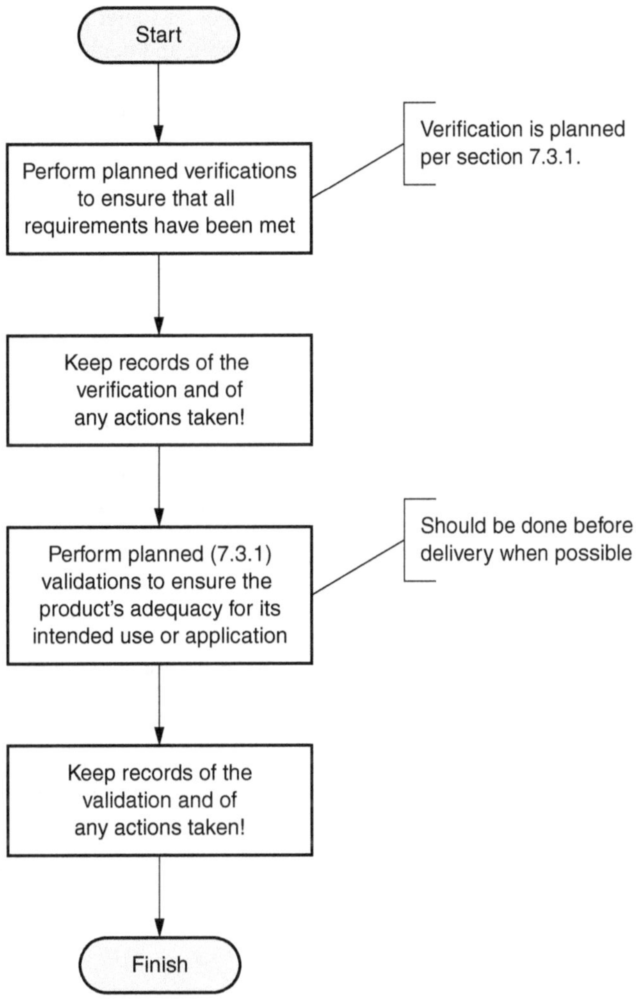

The Standard: 7.0 Product Realization

7.3 Design and Development

7.3.7 Design and Development Validation

13485	The organization must validate a product's ability to meet customer specifications and "known intended use or application" at the conclusion of the design and development process. Validation is to be conducted before product delivery or its "implementation" whenever it is possible to do so after the product's completion. Validation results and any necessary actions subsequent to the validation are to be treated as quality records.
9000	Defines *validation* in 3.8.5 as "confirmation, through . . . objective evidence, that the requirements of a specific intended use or application have been fulfilled."

Document Requirements:

Record (in addition, a procedure, while not required, is encouraged).

Internal Audit Questions:

- Is design and/or development validation performed to confirm that the product is capable of meeting requirements for intended use?
- Is validation completed prior to delivery, when applicable?
- Is partial validation provided when full validation cannot be performed prior to delivery?
- Are design validation results documented?
- Are follow-up actions recorded?

Management Summary:

- Customers, suppliers, and people in the organization should be able to use and evaluate the validation output prior to:
  - Construction, installation, or application of engineering designs
  - Installation or use of software outputs
  - Widespread introduction of services
- Clinical evaluations and customer activities (assembly or installation) must also be considered in validation planning and activities.

7.3 Design and Development

7.3.7 Design and Development Validation

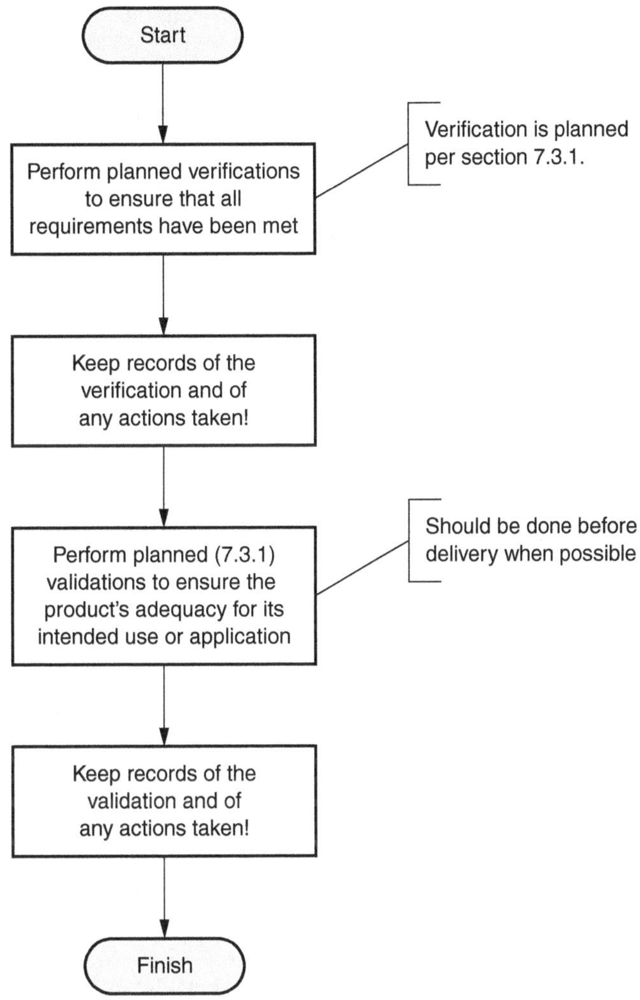

The Standard: 7.0 Product Realization

7.3 Design and Development

7.3.8 Design and Development Transfer

13485 Procedures for design and development transfer are now required

7.3.9 Control of Design and Development Changes

13485 Link to risk management and product realization added. Detail regarding determining significance of a change added.

7.3.10 Design and Development Files

13485 When a change to a product's design and/or development occurs, it must be noted, with the information to be treated as a quality record. The organization must examine each change to ensure that the product will remain in conformance with customer and any other requirements before approving a change. The change's impact on both components and product already delivered to the customer must be covered in the examination.

Document Requirements:

Records:

Required in subsections 7.3.3, 7.3.4, 7.3.5, 7.3.6, 7.3.7, 7.3.8, and 7.3.9.

Remember:

A procedure for the activities in section 7.3 is not required but encouraged.

Internal Audit Questions:

- Are all design and/or development project changes documented?
- Is there evidence to demonstrate that changes are evaluated and authorized prior to the implementation of the change?
- Do records include the results of review of changes (impact, and so on) by the appropriate personnel or responsibilities?
- Have changes been communicated to interested parties?
- Do records include follow-up actions related to the review of changes?

Management Summary:

Verification and validation data should be reviewed by methods such as:

- Process capability
- Process and product improvement
- Investigation of failures
- Evaluation of future design and development process needs
- Recording of results and conclusions of the transfer
- Maintenance of records of changes, their review, and any necessary actions
- Maintenance of a design and development file for each medical device type or medical device family

7.3 Design and Development

7.3.8 Design and Development Transfer

7.3.9 Control of Design and Development Changes

7.3.10 Design and Development Files

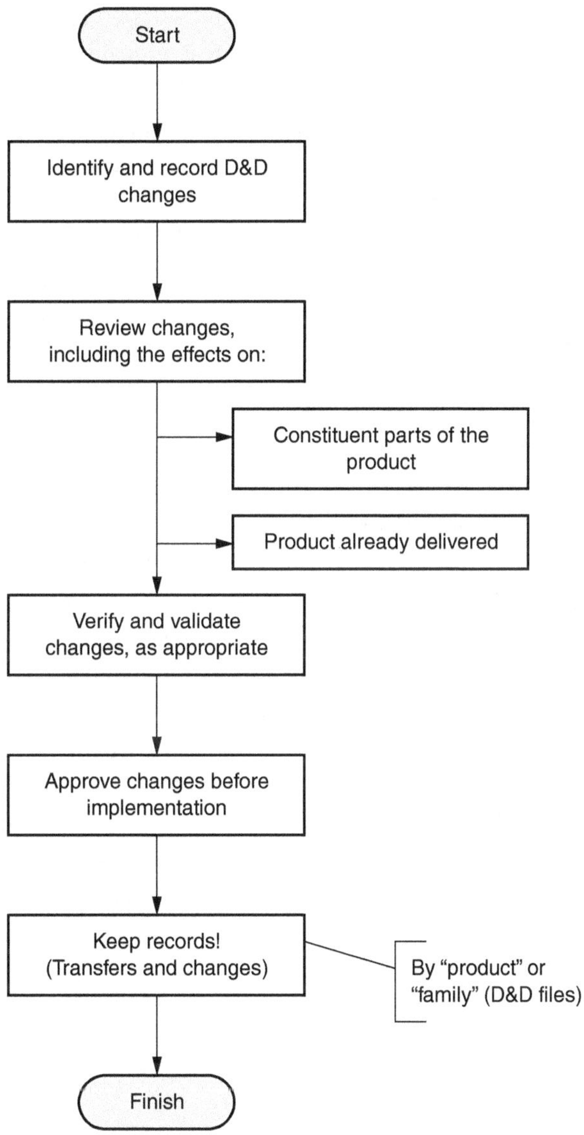

The Standard: 7.0 Product Realization

7.4 Purchasing

7.4.1 Purchasing Process

13485	The organization must make sure that materials, components, and services purchased from suppliers satisfy customer specifications for the organization's product, with the degree of supplier oversight proportional to the impact a part or service has on the product or processes. Suppliers must be qualified and capable of supplying parts/services needed to satisfy product and organizational requirements, based on criteria developed by the organization for choosing suppliers and engaging in ongoing verification of qualifications and capability. The organization must treat as quality records the results of its verification activities and actions taken in response to verification outcomes.
9004	Recommends electronic linkage with suppliers to ensure communication of requirements and supplier participation in setting specifications for parts and services. Encourages consideration of 15 types of activity in the purchasing process. Provides 10 inputs organizations should use in managing supplier quality, and advises organizations to assist suppliers in their development.
9000	Notes in 3.3.6 that suppliers include providers of a service or information.

Document Requirements:

Record

Internal Audit Questions:

- Have criteria for selection, monitoring, and periodic evaluation of suppliers been defined?
- Does a process exist for selecting and evaluating suppliers based on the risk to the organization's intended use for the product or service received?
- Are evaluation results documented and retained as records?
- Are follow-up actions documented?
- Is there a written procedure on how the organization ensures that purchased product meets the applicable requirements?

Management Summary:

- Management should consider electronic communication of requirements to suppliers.
- The organization should consider involving suppliers in the purchasing process to help the organization control inventory.
- Evaluations (supplier performance, auditing, and so on) can drive improvements in suppliers' systems that can in turn drive improvements in the product or organization.
- Quality agreements, or other documentation, can be used to establish expectations between organizations.

7.4 Purchasing

7.4.1 Purchasing Process

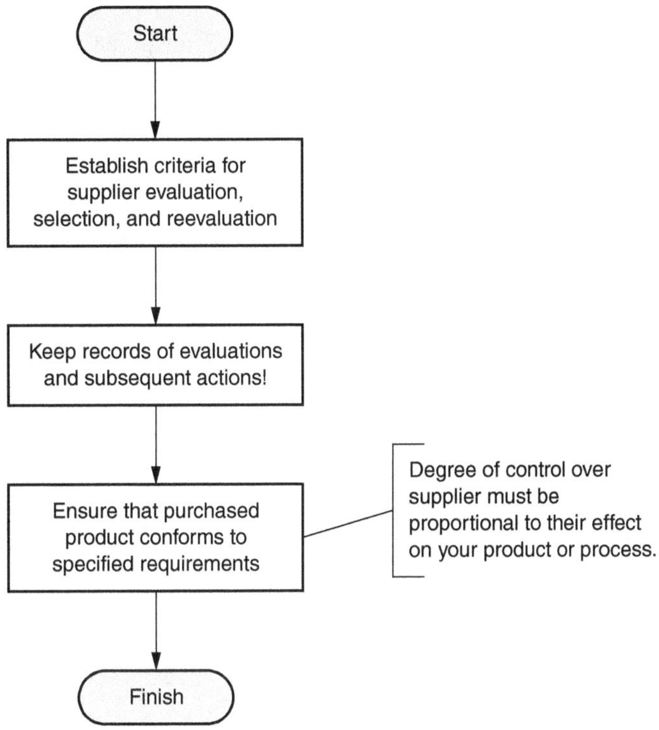

The Standard: 7.0 Product Realization

7.4 Purchasing

7.4.2 Purchasing Information

13485 The specifications for a product being purchased from a supplier must be part of the purchasing information, which must include approval requirements for the product and related items and activities, competency requirements for employees working on the product, and QMS requirements for the supplier's operations. The organization is required to make sure all requirements are identified and spelled out before transmitting the purchasing information to the supplier.

Document Requirements:

Record

Internal Audit Questions:

- Does purchasing information adequately describe the products being ordered?
- Does purchasing information include (where appropriate) requirements for approval or qualification of product, procedures, processes, equipment, and personnel?
- Does purchasing information include (where applicable) quality management system requirements?
- Is purchasing information reviewed and approved to assure adequate descriptions of the specified requirements prior to issuing to the supplier?
- Are there requirements in place to ensure, as applicable, that changes in a provided product or service are communicated prior to implementation?
- Are there requirements in place to assure adequate traceability of purchased materials?

Management Summary:

Management should consider the following when creating purchasing processes:

- Unique supplier processes
- Warranty replacement for nonconforming purchased products
- Logistic requirements
- Risk of the purchased product or service to the organization's processes
- Control of purchased product deviating from requirements
- Access to supplier's premises
- Mitigation of risks associated with purchased product
- Product identification and traceability

7.4 Purchasing

7.4.2 Purchasing Information

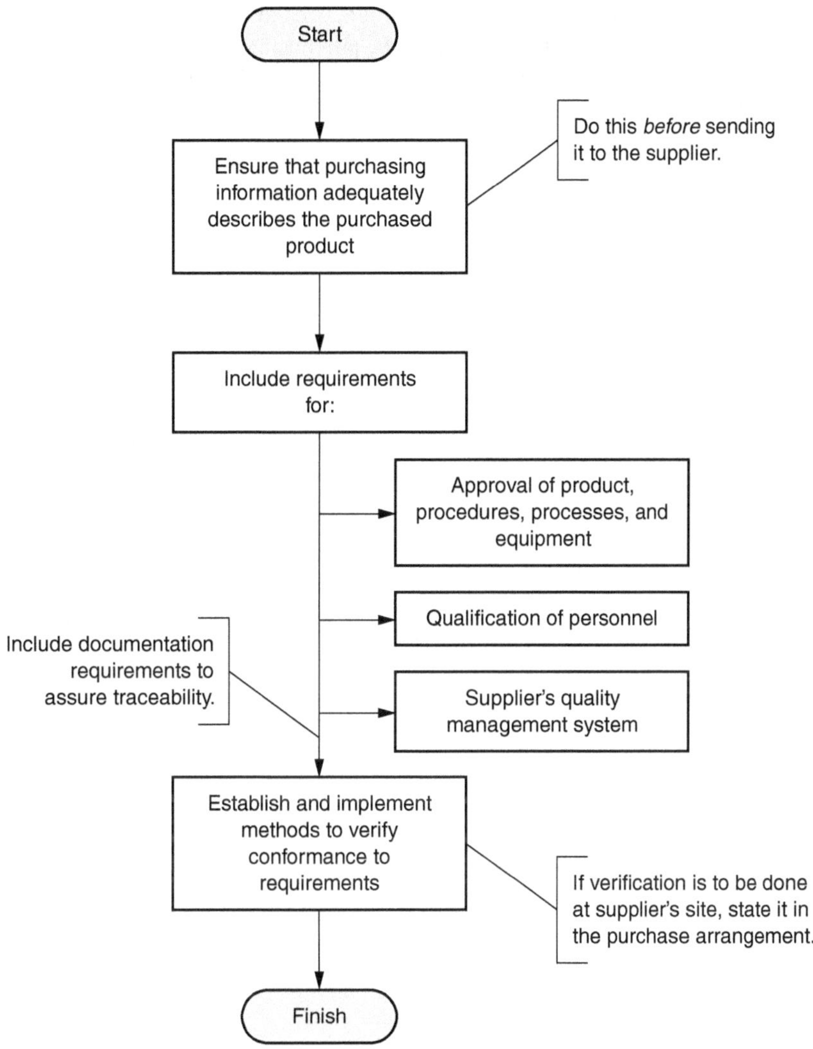

The Standard: 7.0 Product Realization

7.4 Purchasing

7.4.3 Verification of Purchased Product

13485	The organization is to create and use verification practices to confirm that purchased product conforms to the requirements conveyed to the supplier in the purchasing information. The purchasing information also must indicate whether the organization or its customer will conduct on-site inspections of the product, processes, and/or supplier's facility and the related procedures for product release.
9000	Defines *release* in 3.6.9 as "permission to proceed to the next stage of a process."

Document Requirements:

Record

Internal Audit Questions:

- Has the organization defined a process for verifying that purchased product conforms to defined requirements?
- Is the process effectively implemented?
- Does objective evidence exist of product acceptance?
- Is verification of purchased product performed at the supplier's premises?
- If so, are the arrangements specified, and does objective evidence exist to show effective implementation?

Management Summary:

- In the event of supplier failure, management should have actions in place to maintain the organization's performance.
- The criticality of the purchased product, and the level of in-process and final inspection, should be considered when determining the level of monitoring required on incoming product.

7.4 Purchasing

7.4.3 Verification of Purchased Product

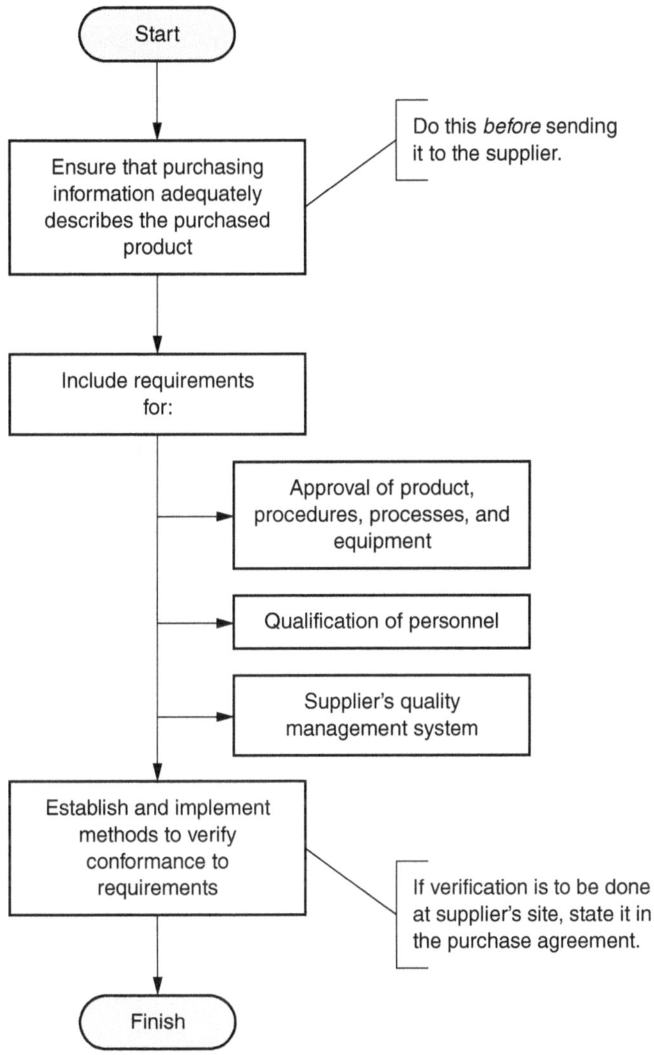

The Standard: 7.0 Product Realization

7.5 Production and Service Provision

7.5.1 Control of Production and Service Provision

7.5.2 Cleanliness of Product

13485	Documented requirements for cleanliness of product or contamination control of product ([a]–[e]).

7.5.3 Installation Activities

13485	Records of medical device installation and verification of installation performed by the organization or its supplier shall be retained.
13485	Production and service operations are to be conducted under planned procedures that control the activities involved, including production, monitoring and measuring, delivery, and post-delivery servicing. To ensure that processes and products conform to customer specifications and other requirements, the organization is required to control six items, including the availability and/or use of procedures, work instructions, information detailing product characteristics, and proper equipment.
9004	Provides guidance on going beyond control of production and service processes to increased effectiveness and efficiency and associated activities that support these processes, with examples of the support activities provided.
9000	Defines *quality control* in 3.2.10 as "part of quality management, focused on fulfilling quality requirements."

Document Requirements:

- Documented procedure
- Record

Internal Audit Questions:	*Management Summary:*
• Are specifications available that define quality characteristic requirements of the product and/or service? • Has the organization demonstrated the suitability of equipment for production and service operations to meet product and/or service specifications? • Are work instructions available and adequate to permit control of the appropriate operations to ensure conformity of product and/or service? • Have the work environment requirements been defined, and are they being met to ensure conformity of the product and/or service? • Is suitable measuring and monitoring equipment available when and where necessary to ensure conformity of the product and/or service? • Have monitoring and verification activities been planned, and are they carried out as required? • Are suitable processes in place for the documentation of acceptance and/or conformance throughout provision of products or services? • If applicable, is there a requirement for the documentation of cleaning or sterilization processes? • Have suitable processes for hardware, processed material, and software been implemented for release of the product and for its delivery to the customer? • Are labeling and packaging operations properly implemented and defined? • Are suitable release mechanisms in place to ensure that a product or service conforms to requirements? • Have processes been defined to ensure traceability of the products produced and the amount released for distribution?	To improve the effectiveness and efficiency of the realization processes, management should consider the following: • Reducing waste. • Developing supplier capability. • Improving infrastructure. • Processing methods and process yield. • Monitoring methods. • Additional processing (that is, cleaning or sterilization). • Testing requirements. • Documentation requirements. • Ensuring that all production and service operations that require control are defined (including those that need ongoing monitoring, or work instructions). • Establishing and maintaining a record (see 4.2.5) for each medical device or batch of medical devices. • Maintaining and making evident documented requirements for cleanliness of product or contamination control of products. • Maintaining records of medical device installation and verification of installation performed by the organization or its supplier.

7.5 Production and Service Provision

7.5.1 Control of Production and Service Provision

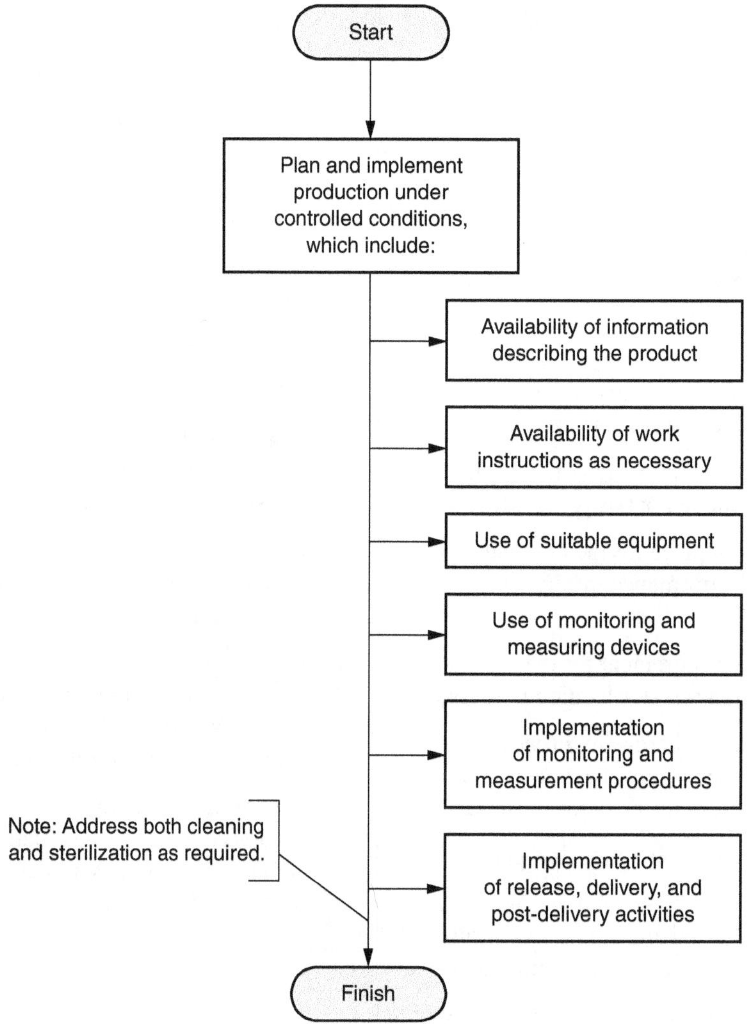

The Standard: 7.0 Product Realization

7.5 Production and Service Provision

7.5.4 Servicing Activities

13485 Servicing activity records must be analyzed to determine whether the issue is a complaint or must be utilized as an improvement input.

7.5.5 Particular Requirements for Sterile Medical Devices

13485 The company will maintain records of the sterilization process parameters used for each sterilization batch.

7.5.6 Validation of Processes for Production and Service Provision

13485 When a product or service cannot be tested during and/or after production and/or service delivery to verify its conformance to customer and other requirements, the organization must ensure that the processes involved are capable of producing a product or service that conforms to requirements, validating the processes. There are six items the organization must have in place to ensure that processes are validated and remain in conformance, including criteria for process review and approval, employee and equipment approval procedures and defined production/service techniques, methods, and processes to be followed. Requirements for revalidation should also be defined.

7.5.7 Particular Requirements for Validation of Processes for Sterilization and Sterile Barrier Systems

13485 Processes for sterilization and sterile barrier systems shall be validated prior to implementation and following product or process changes, as appropriate.

Document Requirements:

- Documented procedure (validation of applicable computer software)
- Record (d)/Record/Record/Record

Internal Audit Questions:

- Are there defined criteria for planning, review, and approval of the validation of processes?
- If applicable, is software validation defined?
- If applicable, is sterilization validation addressed?
- Are the requirements for revalidation defined?
- Is there a process in place to handle deficiencies with product that is in use or that has been delivered?
- Are there records of validation of processes for production and service provision?

Management Summary:

- The organization must also validate processes when product that is in use or has been delivered is determined to be deficient.
- The organization shall analyze records of servicing activities carried out by the organization or its supplier.
- The organization shall maintain records of the sterilization process parameters used for each sterilization batch.
- Records of the results and conclusion of validation and necessary actions resulting from the validation shall be retained.

7.5 Production and Service Provision

7.5.4 Servicing Activities

7.5.5 Particular Requirements for Sterile Medical Devices

7.5.6 Validation of Processes for Production and Service Provision

7.5.7 Particular Requirements for Validation of Processes for Sterilization and Sterile Barrier Systems

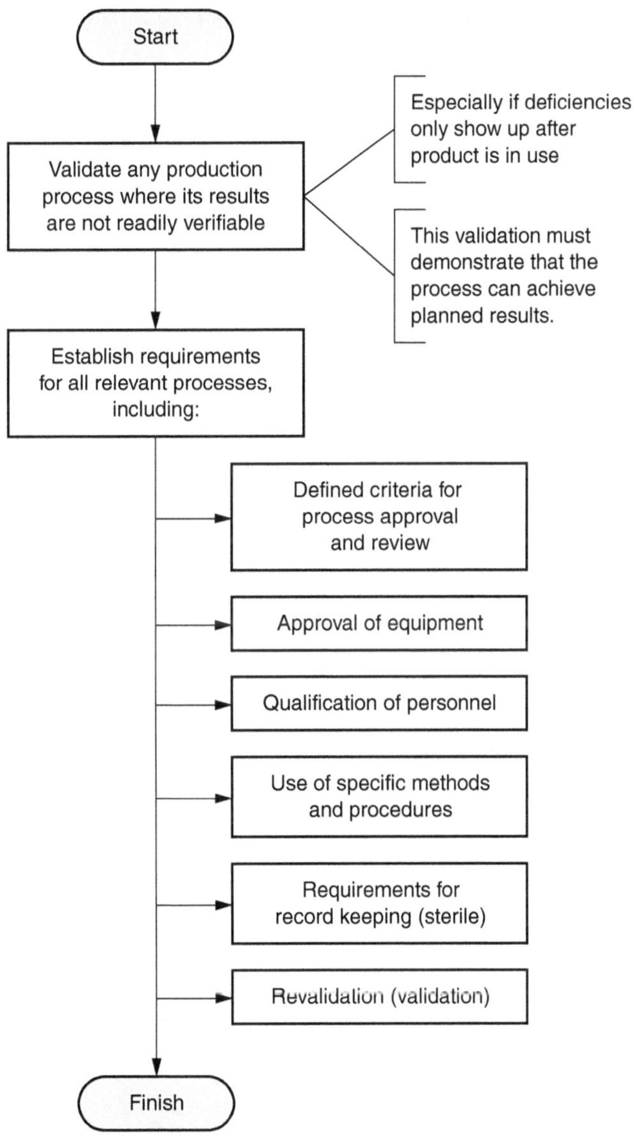

The Standard: 7.0 Product Realization

7.5 Production and Service Provision

7.5.8 Identification

7.5.9 Traceability

13485	The organization must have an effective system in place to maintain identification of a product throughout the production process, including its inspection and test status whenever these are needed. When a product must be traceable, the organization must impose a "unique identification" on each product and record the identification information to permit traceability, with the record to be treated as a quality record.
9004	Recommends collecting data during the identification and traceability processes that go beyond product requirements to improving the product and its processes.
9000	Notes in 3.5.4 that traceability can relate to the origin of materials and parts, among other things.

Document Requirements:

- Documented procedure
- Record

Internal Audit Questions:

- Has the product been identified by suitable means throughout production and service operations?
- Has the status of the product been identified at suitable stages with respect to measurement and monitoring requirements?
- Are there requirements for identification and segregation of nonconforming product?
- For medical devices, is there a written procedure for the identification and handling of returned product?
- Is traceability a requirement? If so, is the unique identification of the product recorded and controlled?

Management Summary:

In order to collect data for improvement, the organization may establish a process for identification and traceability that goes beyond stated minimum requirements.

7.5 Production and Service Provision

7.5.8 Identification

7.5.9 Traceability

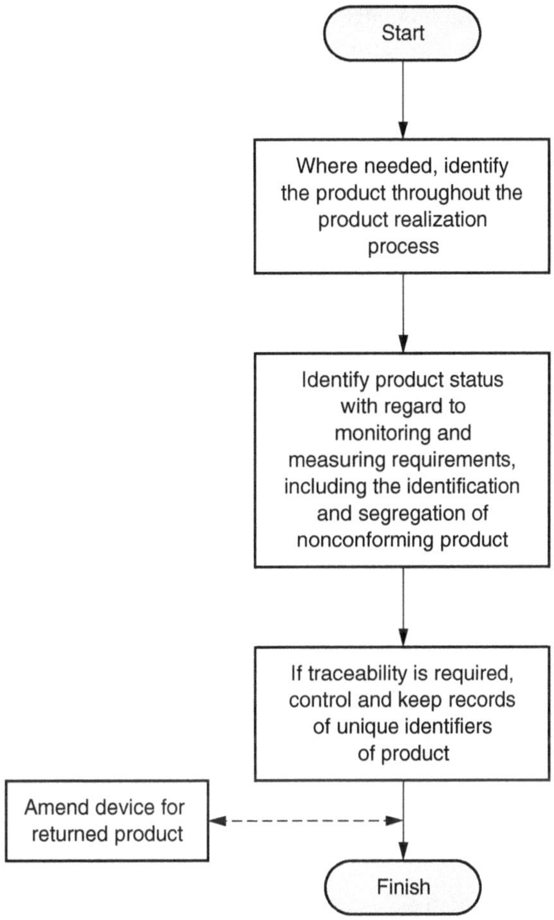

The Standard: 7.0 Product Realization

7.5 Production and Service Provision

7.5.10 Customer Property

13485	The organization must have procedures to protect from damage, misuse, or loss property supplied by the customer for use by the organization in its processes or as components in the product to be supplied to the customer, including intellectual property or confidential data. The organization must alert the customer if its property is nonconforming, and keep records of these incidences as quality records.
9004	Provides examples of the types of property customers or other interested parties might entrust to an organization for use in production or service activities.

Document Requirements:

Record

Internal Audit Questions:

- Has the organization identified, verified, protected, and maintained customer property provided for incorporation into the final product?
- Does control of product extend to all customer property, including intellectual property?
- Does the organization handle confidential health data (complaints, and so on)? If so, how are these data controlled?
- Are there records that indicate when customer property has been lost, damaged, or otherwise found to be unsuitable?
- When customer property has been lost, damaged, or otherwise found to be unsuitable, is there evidence that the customer has been informed?
- Are there defined processes for reporting the destruction of customer property while attempting to maintain preservation of customer provided product or property?

Management Summary:

Examples of customer property that the organization may be responsible for include:

- Ingredients or components for inclusion in a product
- Product supplied for return repair, maintenance, or upgrading
- Packaging materials
- Labeling, including instructions for use
- Customer property to be transported to a third party
- Customer intellectual property
- Confidential data supplied (complaints, investigations, and so on)

7.5 Production and Service Provision

7.5.10 Customer Property

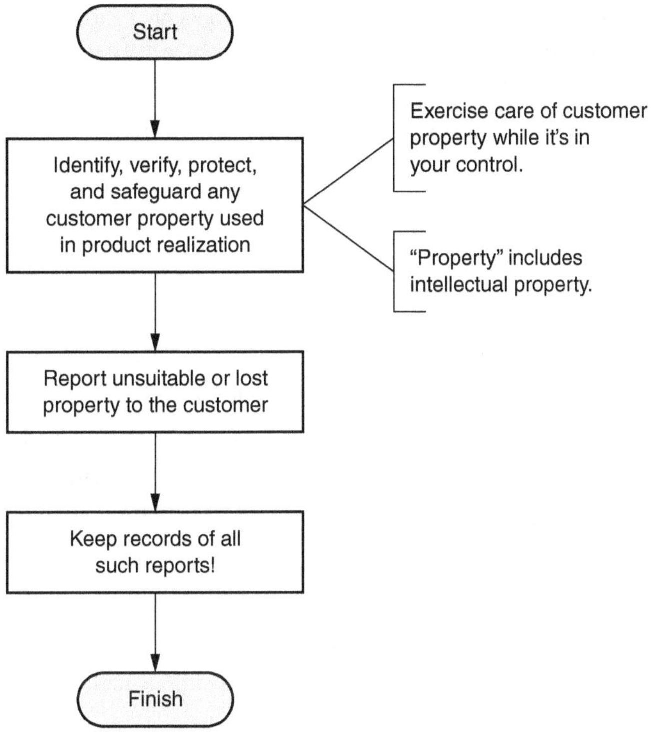

The Standard: 7.0 Product Realization

7.5 Production and Service Provision

7.5.11 Preservation of Product

13485	The organization must ensure that a product and its components continue to satisfy customer specifications and other requirements throughout the production process until delivery to the customer's intended operations.
9004	Advises management to ensure that resources are available to protect a product from destruction, deterioration, or misuse while in the organization's possession, and to inform the customer and others of what is required to preserve the product during its life cycle.

Document Requirements:

Documented procedure

Internal Audit Questions:

- Is there a written procedure addressing the preservation of product?
- Is product identified during internal processing and delivery?
- When handling product during internal processing and delivery, does the organization preserve conformity to customer requirements?
- Are there requirements for investigating nonconformances?
- When packaging product during internal processing and delivery, does the organization preserve conformity to customer requirements?
- When storing product during internal processing and delivery, does the organization preserve conformity to customer requirements?
- Does the organization protect the product during internal processing and delivery to preserve conformity to customer requirements?

Management Summary:

To control handling, packaging, storage, preservation, and delivery of product, management should consider special requirements that may arise from the nature of the product, such as:

- Software
- Electronic media or components
- Hazardous materials
- Products requiring special handing
- Products requiring special storage conditions
- Products requiring special service people
- Unique or irreplaceable products and materials

7.5 Production and Service Provision

7.5.11 Preservation of Product

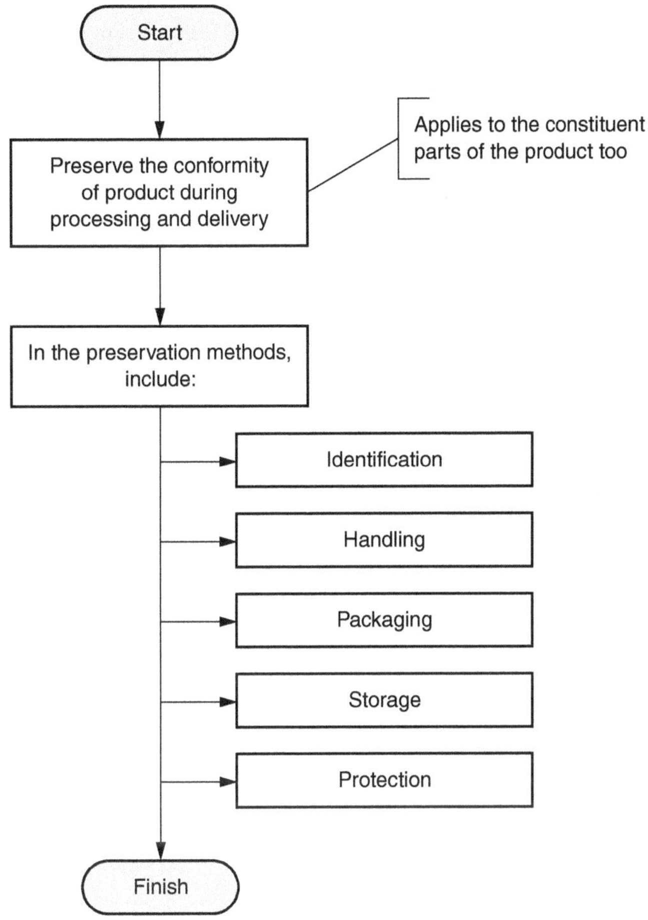

The Standard: 7.0 Product Realization

7.6 Control of Monitoring and Measuring Equipment

13485	The organization is to decide what types and frequency of monitoring and measuring are required to ensure that a product satisfies customer and other requirements, identify the devices necessary for these activities, and put in place procedures to ensure that monitoring and measurement are conducted in accordance with the organization's specifications. The organization must maintain the devices in a condition to ensure accurate monitoring and measurements, including calibration, with the results, along with device maintenance records, to be treated as quality records ([a]–[e]). Specific requirements are spelled out for computer software used for monitoring and measurement purposes. When a device is out of conformity, previous measurements must be examined to ensure that the nonconformity did not result in product nonconformities.
9004	Expands on what monitoring and measurement processes could involve and achieve, with emphasis on the organization taking steps to eliminate potential problems within processes and thus reduce the need to rely on monitoring and measuring activities.
9000	Provides related definitions in 3.8.2, Inspection, and subsection 3.10, Terms related to quality assurance for measurement processes.

Document Requirements:

- Record (a)
- Documented procedure

Internal Audit Questions:	*Management Summary:*
• Is there a written procedure to ensure that required monitoring and measuring are carried out? • Has the organization identified the measurements to be made? Has the organization identified the measurement and monitoring devices required to assure conformity of product to specified requirements? • Are measuring and monitoring devices used to ensure measurement capability? Are they calibrated and adjusted periodically or prior to use against devices traceable to international or national standards? Are those calibration results recorded? • Are processes in place to determine, where possible, the status of equipment prior to maintenance, repair, or calibration activities? • Is the validation status of measuring equipment easily determined? • When traceability to international or national standards cannot be done since no standards exist, is the calibration recorded? • Are measuring and monitoring devices safeguarded from adjustments that would invalidate the calibration? Are they protected from damage and deterioration during handling, maintenance, and storage? • Does the organization have the validity of previous results from measuring and monitoring devices reassessed if they are subsequently found to be out of calibration? Is corrective action taken? • Is software used for the measuring and monitoring of specified requirements validated prior to use?	• For data confidence, measuring and monitoring processes should include confirmation that the device is fit for use, identifiable, and maintained to suitable accuracy and accepted standards. • Identification of the method (and version) used and the status of that method should be considered.

7.6 Control of Monitoring and Measuring Equipment

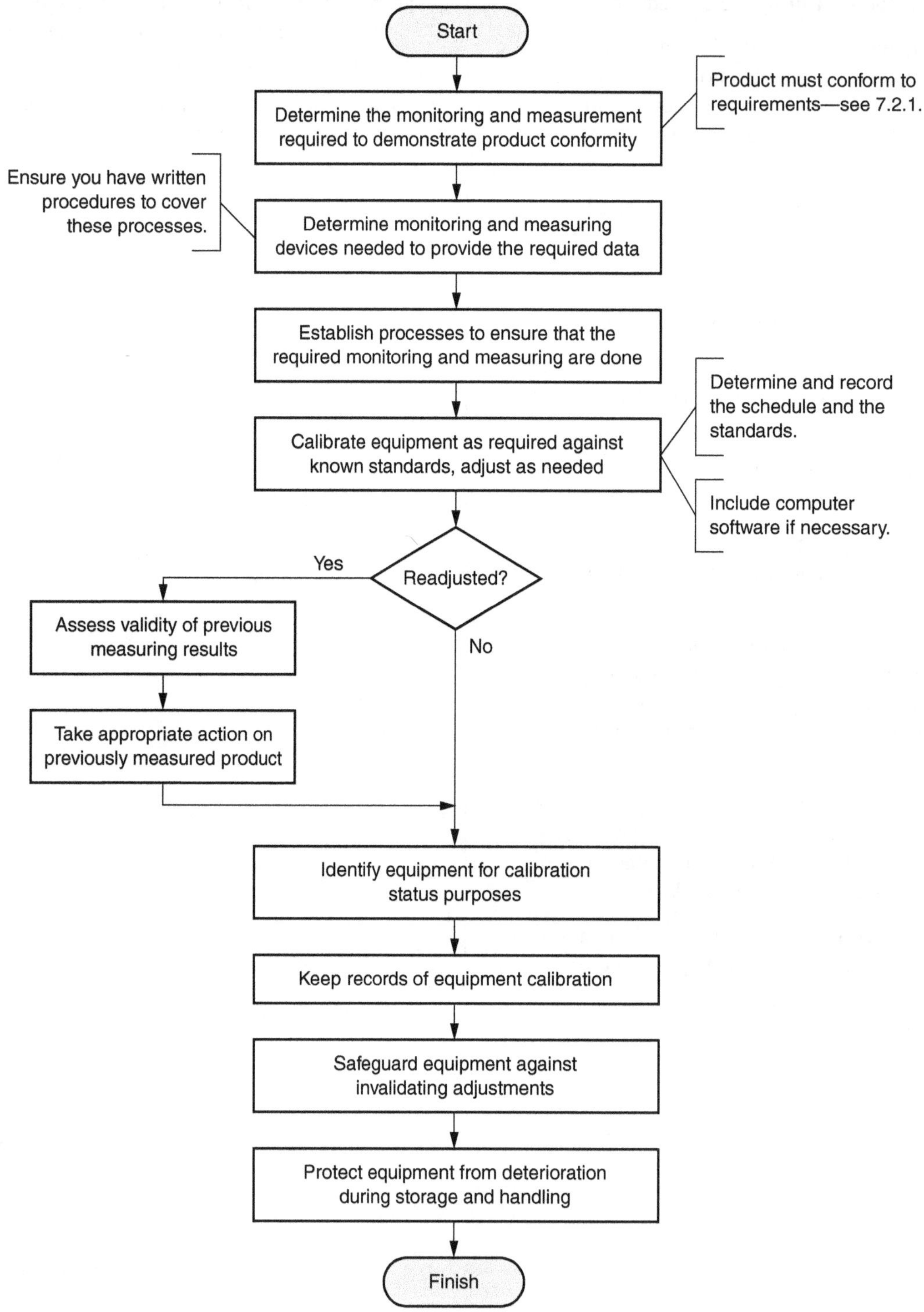

Clause	Change	Interpretation
7.1 (Planning of product realization) (a)	Organizations must document one or more processes for risk management as part of their product realization activities.	Organizations must be able to demonstrate a documented risk management process for the activities they perform, for example, DFMEA and/or PFMEA.
7.1 (Planning of product realization) (b)	Organizations must ensure that sufficient resources are available to manage the work environment and infrastructure.	Organizations must be able to demonstrate that sufficient resources are available to manage their work environment and infrastructure so that product realization processes are not impacted.
7.1 (Planning of product realization) (c)	Handling, storage, distribution, and traceability activities must now be documented, specific to the product.	Organizations are required to have greater emphasis on documented procedures for these areas of the product realization process.
7.2 (7.2.1 Determination of requirements related to product (d))	Organizations must determine and document whether any user training is required in order to demonstrate the safe use and effective performance of the device.	User training may involve training that is performed by the client for its customer(s) and any internal training provided to the client's staff. The organization must document how user training is performed.
7.2.3 (Communication)	Organizations must demonstrate their communication with the regulatory authorities in accordance with applicable regulatory requirements.	Organizations' typical communications with the regulatory authorities may include the communication of FSCAs, FSNs, and Advice Notes in relation to reporting adverse events. It is also beneficial for clients to maintain two-way communication with notified bodies and certifying authorities.
7.3.2 (Design and development planning)	Organizations must plan and control the design and development of product as the design and development of the product progresses.	Organizations must control and maintain documents used in the design planning stages of a device, including any design changes. Clients must also ensure that all documentation related to the product design are inputs into the design history file (DHF) for the product.
7.3.2 (b) (Design and development planning)	Organizations must document the design review process.	Organizations must document all design review meetings, which will become part of the design history file. Review meetings may include formal review of the project plan, meeting minutes, and a roster of those attending the meeting.

Continued

Clause	Change	Interpretation
7.3.2 (e) (Design and development planning)	Organizations must document and demonstrate traceability of design and development outputs to their corresponding inputs.	Organizations must be able to demonstrate the link from design and development outputs to their corresponding inputs. Examples of ways to demonstrate these linkages include document control, change control, validation protocols, and reports.
7.3.2 (f) (Design and development planning)	Design and development planning must include evidence of resource requirements, including the competence levels of the personnel involved in the design process.	Organizations must be able to document the resources needed for the design and development process, for example, testing and manufacturing facilities required, and roles and responsibilities of the team, including the competency of each team member to perform their role.
7.3.3 (a) (Design and development inputs)	Inputs relating to product requirements shall be determined and records maintained. Inputs shall include functional performance, and usability and safety requirements, depending on the intended use.	Organizations must provide data to support product usability, and this can be sought from the end user. Clients could also use similar (predicate) devices, which could themselves be a design input.
7.3.6 (a) (Design and development verification)	Organizations must document design verification plans (including methods, acceptance criteria, and statistical techniques used). Verification must also include confirmation that design outputs meet the inputs when connected or interfaced with other devices.	Organizations must provide further design verification as detailed, and this is key where devices are used in connection with or interfaced with other devices as part of their function and use.
7.3.7 (Design and development validation)	Organizations must document: (a) Acceptance criteria, sampling plans, and statistical techniques within validation protocols and reports. (b) Design validation shall be conducted on representative product. (c) Design validations must also consider devices interfaced with other products as part of their intended use.	Organizations must now document these key components to support the reproducibility and robustness requirements of design validation. Points (a) and (b) may already be performed by some clients; however, (c) drives clients to consider how their devices interface with other products as part of their intended use and therefore to validate that application as part of the design process. Validation shall also be completed prior to release of the product for use by the customer.

Continued

Clause	Change	Interpretation
7.3.8 (Design and development transfer)	The design transfer process from the design center to manufacturing facility must be documented, and it is now required that design and development outputs are suitable for manufacturing and that product generated meets acceptance criteria and also production capability.	Design and development outputs meeting critical product acceptance criteria will be challenged and verified during the manufacturing stages of validation (IQ, OQ, PQ).
7.3.9 (Control of design and development changes)	A documented procedure must be in place for the following: • Processes for controlling design and development. • Evaluating the impact of change on the product realization process. • Assessing change with regard to function, performance, safety, and usability of the device aligning with regulatory requirements.	Organizations will be expected to have a formal and documented change control process that considers the aspects detailed in the adjacent column.
7.3.10 (Design and development files)	Organizations must maintain a design and development file for each medical device family. The file must contain records demonstrating the conformity to the design. The file must also detail all the changes that have taken place with the product and the evaluation of those changes in the design of the product.	Organizations must be able to demonstrate traceability of all device changes and the evaluation of those changes. Organizations must also show the linkage to the manufacturing specifications for the device, demonstrating how the specifications have been derived during the device design.
7.4.1 (Purchasing process)	Organizations are required to address nonfulfillment of purchasing requirements with the supplier, based on the risk associated with the purchased product on the device and regulatory requirements.	Organizations must determine the impact of the nonfulfillment of purchasing requirements on the finished device. It may not be possible to use the material, based on the nonfulfillment of the requirements.
7.4.2 (Purchasing information)	Purchasing information must include a written agreement that the supplier will notify the client of any changes in the purchased product before those changes are implemented.	Organizations must review those changes and the impact to the product prior to using the material in production in order to determine if any preproduction validation/verification must be performed. This will demonstrate that changes made to the purchased product do not impact fit, form, or function of the device. (See also 7.4.3.)

Continued

Clause	Change	Interpretation
7.5.2 (Cleanliness of product)	Organizations must document the requirements related to the cleanliness of product and contamination control where product cannot be cleaned before it is sterilized and where its cleanliness is a key factor in its end use.	This clarifies the expectation for clients to document how they maintain cleanliness and contamination control during the product realization process so that product cannot be impacted.
7.5.4 (Servicing activities)	Servicing and preventive maintenance data are required to be analyzed for these activities performed by the client or if these activities are subcontracted. These data can then be used as an input into the improvement process.	Organizations need to measure these data, which can serve as input into management review or improvement activities. Satisfactory completion of these activities can also help to determine a level of risk as to whether product is impacted.
7.5.6 (Validation of processes for production and service provision)	Key aspects for validation must now include: • Documenting acceptance criteria. • Rationale for revalidation, if required. • Rationales for statistical techniques and sample sizes adopted. • Changes to validations must now be reviewed and approved with an appropriate level of authority. • Documenting the risk impact of software used and impact to potential product. • Maintaining records of validation results, conclusions, and any outstanding actions from validations.	Organizations are now required to provide clarity for these aspects within validations. Emphasis is placed on clients regarding the control of changes to validation and also that any changes are given appropriate review and approval by the correct level of authority. The impact of any software used is also required to be assessed with respect to impact to product. Organizations must also now maintain records of validation results, conclusions, and any outstanding actions from validations.
7.5.8 (Identification)	Organizations must now: • Identify product status with respect to monitoring and measurement throughout product realization. • Product status is required to be identified so that only product that has met the required acceptance criteria or released under concession is shipped.	Organizations must ensure that identification is performed throughout the product realization process. This applies to the identification of monitoring and measuring devices (also aids to calibration) and test devices. Identification also applies to product status, so that only identified product is used during the product realization process, for example, warehousing, manufacturing, and installation of equipment. Unique device identification (UDI) is applicable to the classification of the device and to meeting regulatory requirements.

Continued

Clause	Change	Interpretation
7.5.9.2 (Traceability)	Device records must now demonstrate safety and performance of the device. The term *agent* has been replaced with the term *supplier of distribution services.*	Organizations must now show that the device performs safely and meets its intended use. This can be shown by the compliance of the product with its required specification from the manufacturing process. In addition, the term *agent* has been replaced with *supplier of distribution services*. This applies to stakeholders who provide a service that impacts the client's QMS, and also includes outsourced services.
7.6 (Control of monitoring and measuring equipment)	Adjustments and readjustments to testing and measuring equipment must be recorded. Greater emphasis on records of calibration and verification are to be maintained. Computer software and its application must now be validated and risk assessed based on its application and use within product realization.	This new change/update requires that clients have to document any adjustments made to testing and measuring equipment, with a brief rationale on the record as to why the adjustment was necessary. There is also a reiteration to assess the impact of computer software used in the product realization process, based on its risk to the product.

REQUIREMENTS REMOVED

Clause	Change	Interpretation
7.1 (Planning of product realization)	*Notes* 1 and 2, relating to *quality plan* and *development of product realization processes* have been removed.	*Notes* 1 and 2 are now included in the main body of text in clause 7.1 of the 2016 version of the standard.
7.3.2 (Design and development planning)	The corresponding section from the 2003 version is clause 7.3.1. The *Note* relating to *design transfer* activities being verified before becoming final production specifications has been removed.	The *Note* is now covered in section 7.3.2 (e) in relation to the traceability of design outputs corresponding to their design inputs.
7.3.4 (Design and development outputs)	The corresponding section from the 2003 version is clause 7.3.3. The *Note* in this section has been removed.	The *Note* referred to specific types of records used to document design and development outputs. This is now covered by a generic statement.
7.3.7 (Design and development validation)	The corresponding section from the 2003 version is clause 7.3.6. *Notes* 1 and 2 have been removed.	The *Notes* are now included in the main body of text in clause 7.3.7 of the 2016 version of the standard.

SUMMARY

by S. Pletzer

ISO 13485:2016—7 Product Realization

7.1 Planning of product realization

The organization plans and develops processes for product realization, including objectives, requirements, verification, validation, monitoring, inspection, testing, and risk management, and provides evidence of meeting all requirements.

7.2 Customer-related processes

The organization determines customer product requirements, reviews such requirements, and communicates with the customer on all product-related issues.

7.3 Design and development

The organization establishes complete design and development procedures, including planning, inputs, outputs, review, verification, validation, and changes.

7.4 Purchasing

The organization establishes purchasing processes to ensure that product meets requirements.

7.5 Production and service provision

The organization provides any product and related services under controlled conditions, including cleanliness, contamination, installation, servicing, sterilization, validation, identification, traceability, customer property, and preservation of product.

7.6 Control of monitoring and measuring equipment

The organization determines monitoring and measuring to be performed, the equipment needed to show evidence of product meeting requirements, and any calibration associated with measurement equipment.

SECTION 8: MEASUREMENT, ANALYSIS AND IMPROVEMENT

8.1 General
8.2 Monitoring and measurement
  8.2.1 Feedback
  8.2.2 Complaint handling
  8.2.3 Reporting to regulatory authorities
  8.2.4 Internal audit
  8.2.5 Monitoring and measurement of processes
  8.2.6 Monitoring and measurement of product
8.3 Control of nonconforming product
  8.3.1 General
  8.3.2 Actions in response to nonconforming product detected before delivery
  8.3.3 Actions in response to nonconforming product detected after delivery
  8.3.4 Rework
8.4 Analysis of data
8.5 Improvement
  8.5.1 General
  8.5.2 Corrective action
  8.5.3 Preventive action

The Standard: 8.0 Measurement, Analysis and Improvement

8.1 General

13485 The organization is to plan for, establish, and improve processes to monitor, measure, and analyze its QMS and procedures, which must provide evidence that its products meet customer and other requirements, verify the QMS's effectiveness, and use and achieve continual improvement of the QMS. This will require identification of what statistical and other techniques are needed and how much they need to be used.

9004 Provides 11 topics to be evaluated in establishing measurement, analysis, and improvement processes. Offers guidance on applying measurement to improve the organization's performance into the future and to assess the value of the measurements in meeting its needs.

9000 Explores in 2.10, Role of statistical techniques, their use in understanding variability to help improve effectiveness and efficiency.

Document Requirements:

None

Internal Audit Questions:

- Is there objective evidence to demonstrate that the organization has defined, planned, and implemented the measurement and monitoring activities needed to demonstrate conformity of product?
- Is there objective evidence to demonstrate that the organization has defined, planned, and implemented the measurement and monitoring activities needed to assure the effectiveness of and to achieve improvement of the QMS?
- Has the organization defined the sources of inputs to and the outputs of these improvement activities?
- Is there objective evidence to demonstrate that the organization has determined the need for and use of applicable methodologies, including statistical techniques (that is, specifications, procedures, work instructions, control plans, process sheets, and so on)?

Management Summary:

- Results of data analysis from improvement activities should be a management review input.
- Management should note that measurements of customer satisfaction are vital for evaluating not only the product's but the organization's performance.

8.1 General

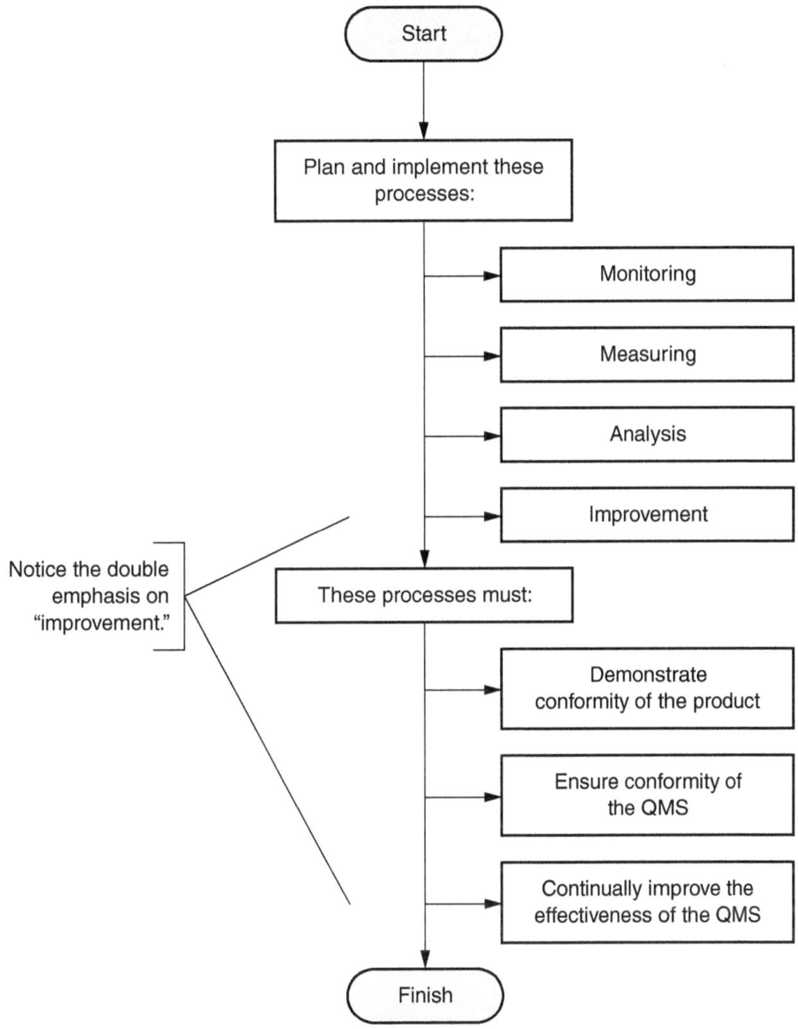
Start
Plan and implement these processes:
Monitoring
Measuring
Analysis
Improvement
Notice the double emphasis on "improvement."
These processes must:
Demonstrate conformity of the product
Ensure conformity of the QMS
Continually improve the effectiveness of the QMS
Finish

The Standard: 8.0 Measurement, Analysis and Improvement

8.2 Monitoring and Measurement

8.2.1 Feedback

8.2.2 Complaint Handling

8.2.3 Reporting to Regulatory Authorities

13485	The organization must establish a program to collect and measure data (for example, round table discussions, surveys, web invites, mailings, follow-up telephone calls) and take action on data regarding customer satisfaction.
9004	Provides guidance on developing a system to measure customer satisfaction, including examples of available information relating to customers and their interaction with the organization, and of sources of information on customer satisfaction.
9000	Defines customer satisfaction in 3.1.4 as the customer's perception of how well its "requirements have been fulfilled."

Document Requirements:

- Documented procedure
- Record

Internal Audit Questions:

- Are customer satisfaction and/or customer dissatisfaction information monitored?
- Are methods for gathering and using customer information determined and deployed throughout the organization?
- Has the customer information been processed in a way to provide quantifiable data and trends?
- Has the organization identified outputs of the results of this processing?
- Is there evidence that the results/outputs are used for improving product realization or processes?

Management Summary:

- The organization should use customer feedback to plan processes that listen effectively and efficiently to the voice of the customer.
- The organization should identify sources of customer and end user information (from internal and external sources) such as:
  - Customer complaints
  - Direct communication with customers
  - Questionnaires and surveys
  - Focus groups
  - Consumer reports
  - Various media reports (articles, regulatory actions, and so on)
  - Sector and industry studies

8.2 Monitoring and Measurement

8.2.1 Feedback

8.2.2 Complaint Handling

8.2.3 Reporting to Regulatory Authorities

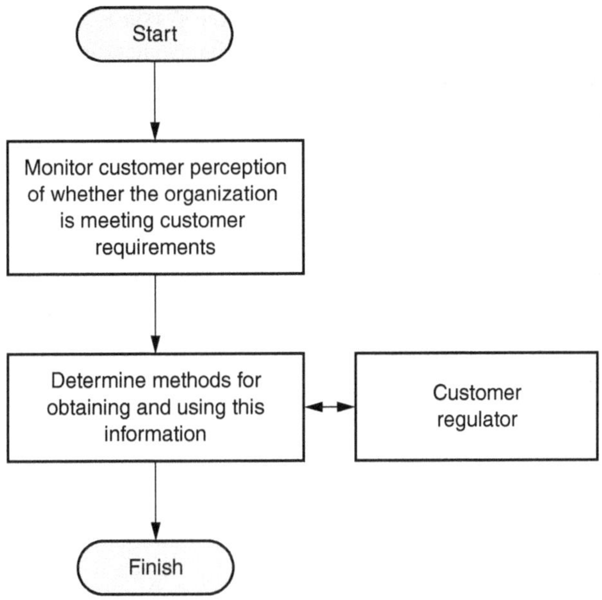

The Standard: 8.0 Measurement, Analysis and Improvement

8.2 Monitoring and Measurement

8.2.4 Internal Audit

13485	A documented internal audit program must be established to assess the QMS's effectiveness in meeting the requirements of ISO 13485 and any other requirements the organization has specified for its QMS, and whether its QMS is functioning and is being conformed to by employees in the course of organizational activities. The organization must plan out and conduct audits of each process, procedure, and area frequently and thoroughly enough to ensure the system's effective operation. Auditors must be capable of accurately assessing the areas they audit. Management for an area where a nonconformity is identified is responsible for ensuring that prompt and effective corrective action is taken, with the internal auditors responsible for verifying that the action eliminates the nonconformity and its causes.
9004	Provides guidance on the purpose and management of internal audits and emphasizes the need for top management to take action to improve the QMS in response to internal audit findings. Offers 11 examples of topics to be examined in internal audits.
9000	Defines 14 terms in 3.9, Terms relating to audit, which may now be found in the revised ISO 19011:2018 standard. Explores role and types of audits in 2.8.2, Auditing the quality management system.

Document Requirements:

- Documented procedure
- Record

Internal Audit Questions:	*Management Summary:*
• Does the organization conduct periodic audits of the quality management system? • Has the organization established an internal audit schedule? • Has the organization demonstrated progress against the schedule? • Do the periodic audits evaluate the conformity of the quality management system to the requirements of ISO? • Do the periodic audits evaluate the degree to which the quality management system has been effectively implemented and maintained? • When planning the audit program, does the organization consider the status and importance of areas to be audited, and results of previous audits? • Are the methodologies, audit scope, and frequency defined? • Do personnel other than those who perform the activity being audited perform the audits? • Is there a documented procedure that includes the responsibilities and requirements for planning and conducting audits? • Is there a documented procedure for recording audit results and reporting to management? • Is there evidence that management takes timely corrective action on deficiencies found during the audit? • Do follow-up actions include the verification of the implementation of corrective action? • Do follow-up actions include the reporting of verification results? • Is there evidence that the results of internal audits are presented to upper management?	• The internal audit process is a management tool for independent assessment of a process or activity. • Planning for internal audits should be flexible in order to allow changes in emphasis based on findings and objective evidence collected during the audit. • Management should provide opportunities for recognition of areas that achieve excellent performance during an internal audit. • Management must show the utilization of audit outputs in the evaluation and improvement of the management system.

8.2 Monitoring and Measurement

8.2.4 Internal Audit

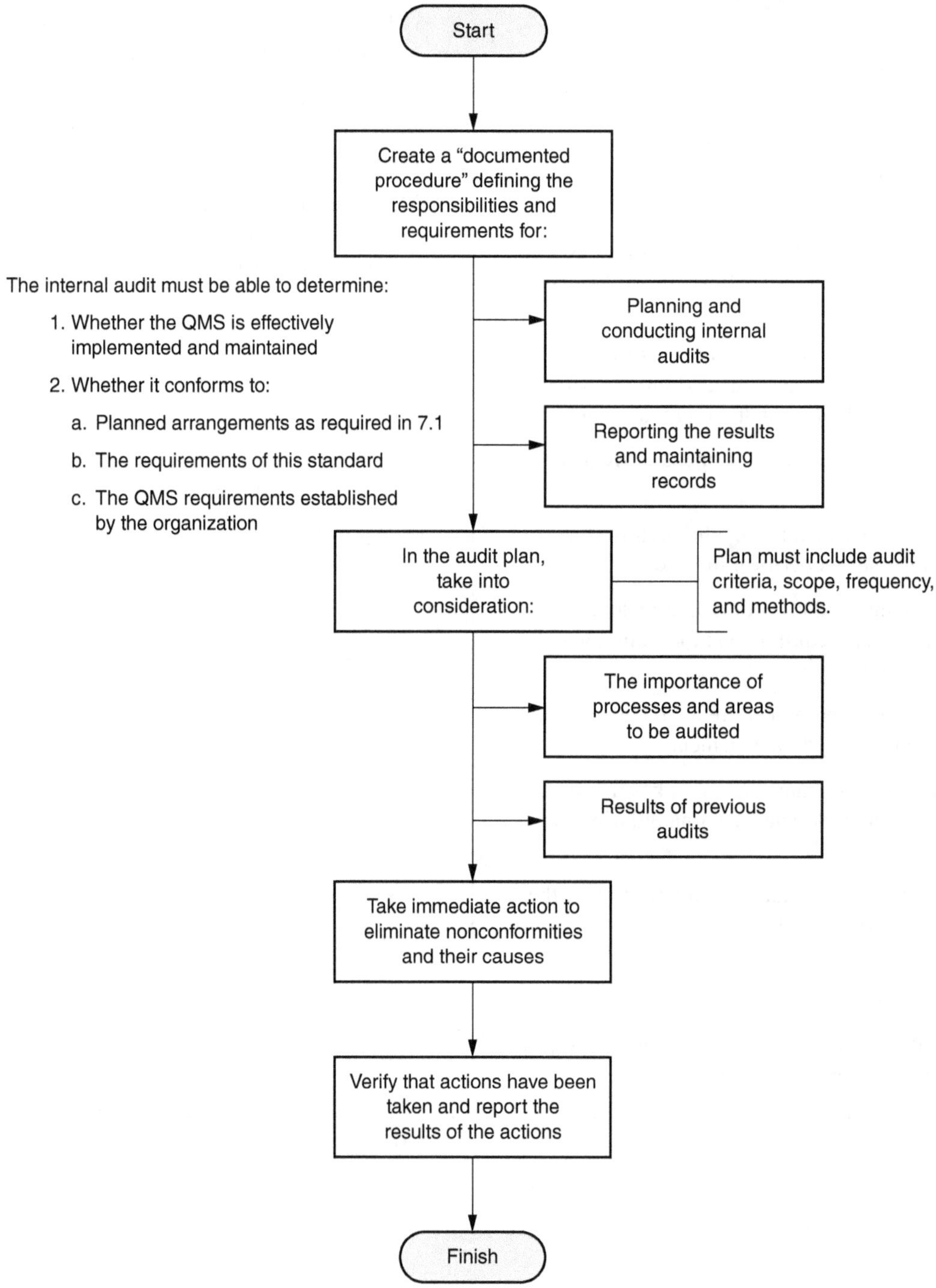

The Standard: 8.0 Measurement, Analysis and Improvement

8.2 Monitoring and Measurement

8.2.5 Monitoring and Measurement of Processes

13485	QMS processes must be monitored and measured to the degree necessary to ensure that the processes produce conforming product and achieve expected outcomes, with corrective actions to be taken when the processes do not produce expected outcomes.
9004	Provides guidance on the use and advantages of self-assessments (including Annex A, Guidelines for Self-Assessment) to improve performance and on the use of QMS information to measure financial performance. Recommends measuring process performance and includes examples of performance aspects that can be measured.
9000	Explores in 2.8.1, Evaluating processes within the quality management system, the questions to be addressed in evaluating a QMS.

Document Requirements:

Record

Internal Audit Questions:

- Has the organization identified the key processes necessary to meet customer requirements?
- Has the organization employed suitable methods to measure and monitor key processes?
- Are the intended purposes of the key processes quantified by process parameter specifications, by specifications for the product output of the process, or by some other means?
- Are the measurement and monitoring methods for realization processes adequate for confirming the continuing suitability of each process to satisfy its intended purpose?

Management Summary:

- The organization should incorporate measurements of process performance into processes and use the measurements in process management.
- Measurements of process performance include:
  - Cycle time or throughput
  - Deviation/CAPA rates
  - Right first time (record review, complaint responses, and so on)
  - Dependability
  - Rework
  - Yield
  - Technology utilization
  - Waste reduction
  - Cost allocation and reduction

8.2 Monitoring and Measurement

8.2.5 Monitoring and Measurement of Processes

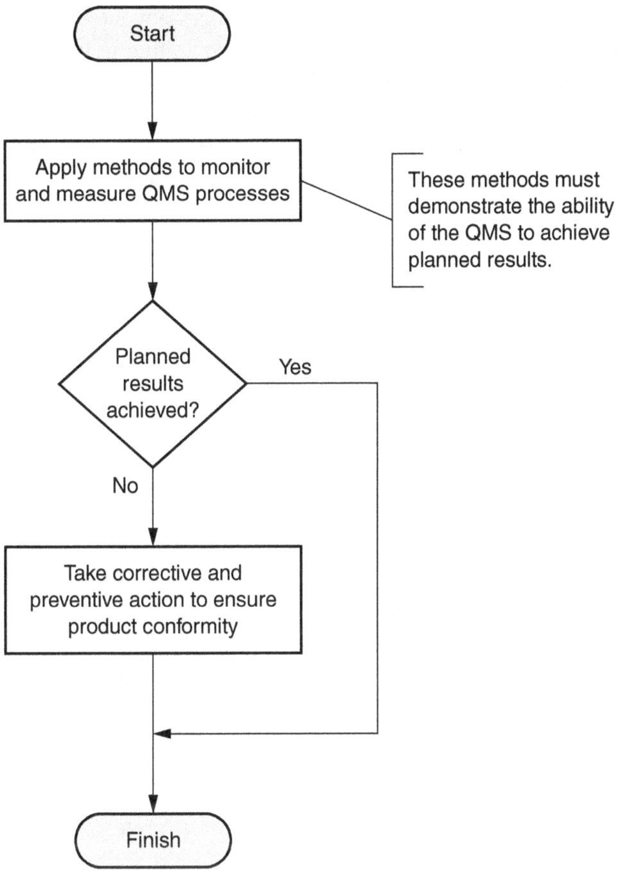

The Standard: 8.0 Measurement, Analysis and Improvement

8.2 Monitoring and Measurement

8.2.6 Monitoring and Measurement of Product

13485	The organization must engage in monitoring and measurement activities throughout its production processes to verify that product conforms to customer and other requirements, maintain its evidence of conformance, and treat documentation identifying who is responsible for a product's release as quality records. A product must not be released until monitoring and measurement indicate that the product conforms to requirements unless the employee responsible for its release authorizes the release beforehand—and the action is acceptable to the customer. The same requirements apply to service delivery.
9004	Recommends using product measurement to improve production processes. Offers 10 aspects and elements of the measurement process that should be considered in selecting measurement methods, and four examples of measurement records to be kept.
9000	Defines *release* in 3.6.9 as "permission to proceed to the next stage of a process."

Document Requirements:

Record

Internal Audit Questions:

- Does the organization measure and monitor product characteristics to verify that product requirements are met?
- Does the organization measure and monitor product characteristics at appropriate stages of the product realization process?
- Is there objective evidence that acceptance criteria for the product have been met?
- Is there identification of the authority responsible for release of the product?
- Are all specified activities performed prior to product release and service delivery?
- When specified activities have not been performed prior to product release or service delivery, is the customer informed?
- Has the customer approved of the action?

Management Summary:

- To improve performance, the organization should review verification records and the methods used for measuring products.
- Measurements of product performance include:
  - Complaint rates
  - Return rates
  - In-process inspection results
  - Nonconformities
  - Yield
  - Sales

8.2 Monitoring and Measurement

8.2.6 Monitoring and Measurement of Product

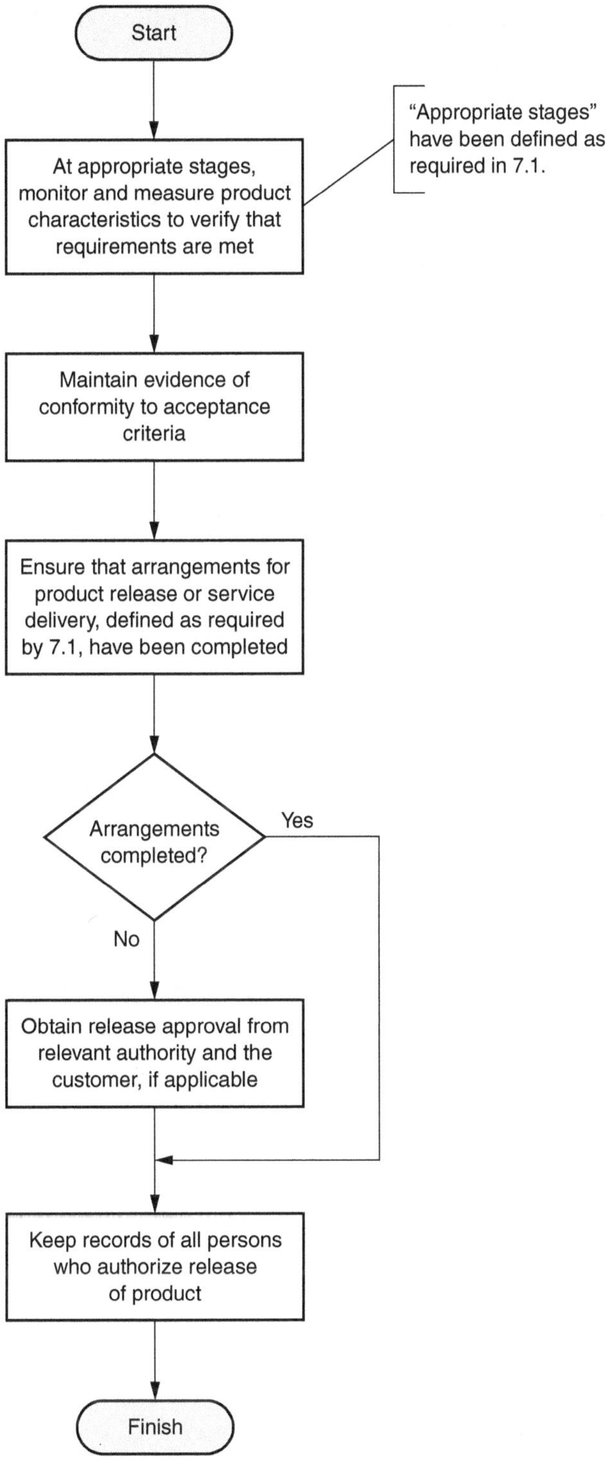

The Standard: 8.0 Measurement, Analysis and Improvement

8.3 Control of Nonconforming Product

8.3.1 General

8.3.2 Actions in Response to Nonconforming Product Detected before Delivery ([a]–[c])

8.3.3 Actions in Response to Nonconforming Product Detected after Delivery

8.3.4 Rework

13485 The organization must prevent unintended delivery or use of nonconforming product, with documentation of the processes to be followed and the authorities responsible for identifying, controlling, and disposing of nonconforming product. The organization must correct nonconformity, offer the product on an as-is basis, or scrap it, with information on the nonconformity and the product's disposition to be treated as a quality record. Corrected product must be reevaluated to ensure that the nonconformity has been eliminated, and the organization is responsible for resolving known or potential nonconformities that arise after a product is delivered or in use.

9004 Examines how top management can establish procedures to deal with nonconforming product that will not only prevent unintended use but will provide feedback to the organization for quality planning and process improvement.

9000 Defines terms relating to the disposition of nonconforming product in 3.6.8, Release, 3.6.10, Repair, 3.6.11, Rework, 3.6.12, Re-grade, and 3.6.13, Scrap.

Document Requirements:

- Documented procedure
- Record/record/record/record

Internal Audit Questions:

- Is there a documented procedure to assure that product that does not conform to requirements is identified and controlled to prevent unintended use or delivery?
- Is there a documented procedure for the review, approval, and acceptance of any reworked product?
- Is there evidence of appropriate action being taken when nonconforming product has been detected after delivery or use has started?
- Is it required that any proposed rectification of nonconforming product be reported for concession to the customer, the end user, or a regulatory body?
- Is it required that the identity of the person(s) authorizing the concession be documented and maintained?
- Is there objective evidence of appropriate communication with the customer when the organization proposes rectification of nonconforming product?

Management Summary:

- To provide analysis data and improvement activities, the organization should record nonconformities to both product realization and support processes.
- During reviews of nonconformities, negative trends should be considered for improvement as well as input to management review for consideration.
- Acceptance of nonconformity disposition may be a contractual requirement of the customer.
- Management should consider not only the stated product requirements but also any details with respect to controls, concessions, and records. This clause has been restructured to be more inclusive.

8.3 Control of Nonconforming Product

8.3.1 General

8.3.2 Actions in Response to Nonconforming Product Detected before Delivery ([a]–[c])

8.3.3 Actions in Response to Nonconforming Product Detected after Delivery

8.3.4 Rework

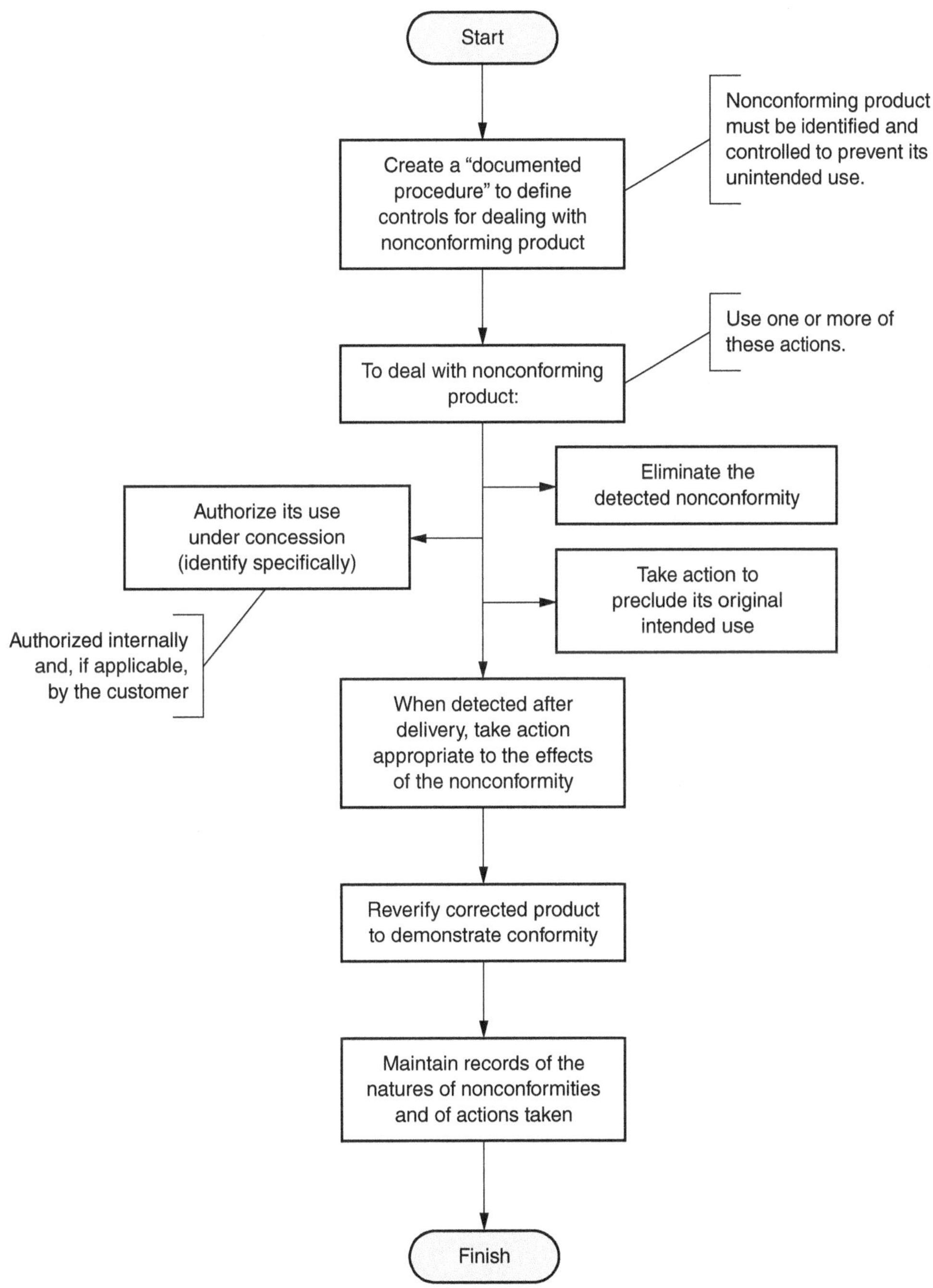

The Standard: 8.0 Measurement, Analysis and Improvement

8.4 Analysis of Data

13485	The organization must use data from its monitoring and measuring activities and other sources to verify that the QMS is appropriate to the organization's needs and effectively conforms to ISO 13485 and other requirements, and to identify opportunities for improvement of the QMS. The standard requires the resulting analysis to provide data on four subjects relating to the QMS and the organization's products.
9004	Stresses the value of fact-based measurements and of identifying root causes of problems. Identifies nine purposes to which analysis can be put.

Document Requirements:

- Documented procedure (covering the activities for 8.4[a]–8.4[f]).
- Record

Internal Audit Questions:	*Management Summary:*
• Is there a written procedure that describes the analysis of data within the organization? • Has the organization determined the appropriate data to be collected to determine the effectiveness of the quality management system and where improvements can be made? • Does the organization analyze appropriate data to determine the suitability and effectiveness of the quality management system and improvements that can be made? • Does the organization analyze appropriate data to provide information on customer satisfaction and/or dissatisfaction and conformance to customer requirements? • Does the organization analyze appropriate data to provide information on characteristics of processes, product, and their trends? • Does the organization analyze appropriate data to provide information on suppliers?	The organization should analyze data to assess performance against plans, objectives, and defined goals, and to identify areas for improvement.

8.4 Analysis of Data

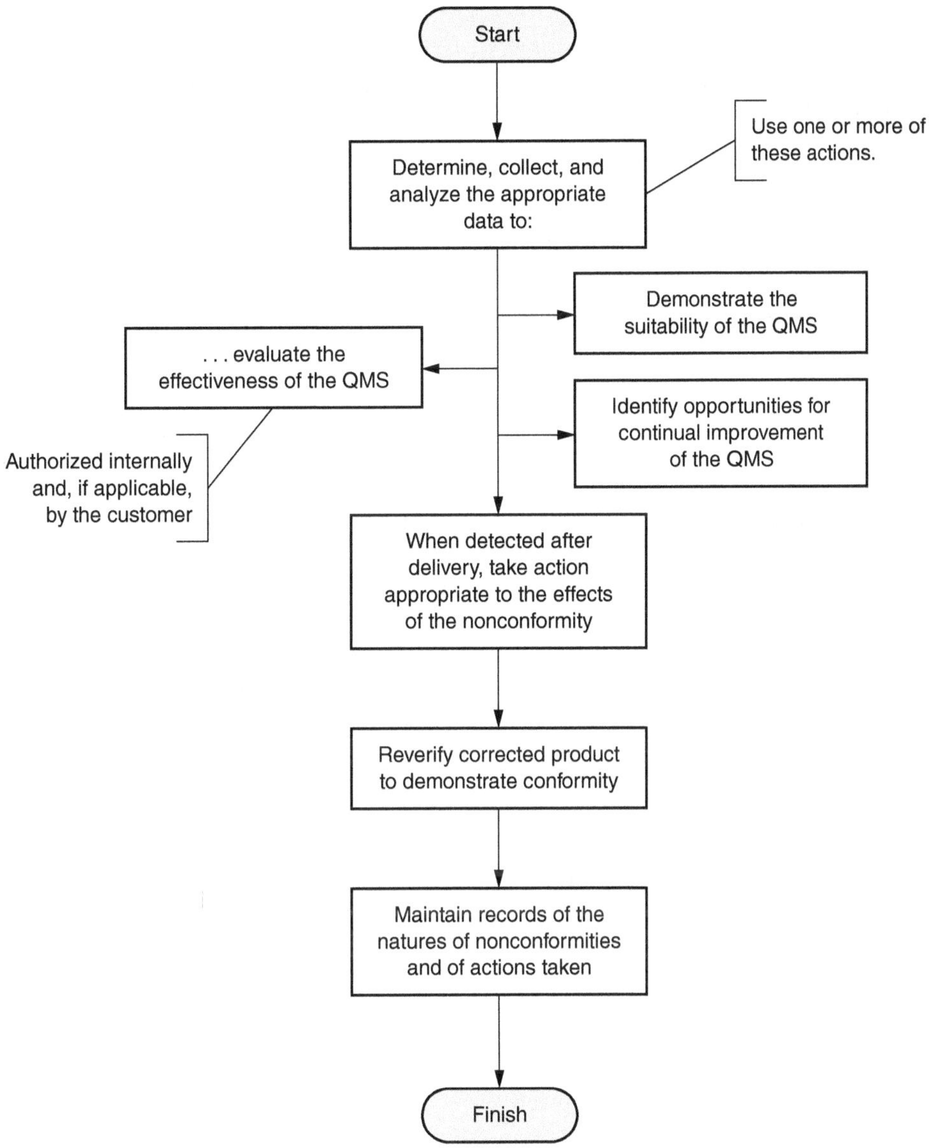

The Standard: 8.0 Measurement, Analysis and Improvement

8.5 Improvement

8.5.1 General

13485	The organization is required to use seven specified QMS elements/processes to "continually improve the effectiveness" of the QMS.
9004	Explores the range of improvements management can pursue, and encourages a proactive approach.
9000	Defines *continual improvement* in 3.1.13 as "recurring activity to increase the ability to fulfill requirements," and explores in 2.9, Continual improvement, the purpose of which is to improve the organization, actions that lead to continual improvement.

Document Requirements:

- Documented procedure
- Record

Internal Audit Questions:

- Is there a written procedure for improvement processes?
- Does the organization plan and manage processes necessary for continual improvement of the quality management system?
- Does the organization use the quality policy, quality objectives, and analysis of data to facilitate continual improvement of the quality management system?
- Does the organization use audit results, corrective action, and preventive action to facilitate continual improvement of the quality management system?
- Does the review of quality data result in action items designed to improve the quality management system (also see Section 5.6.3)?
- Is there a written procedure that addresses the review and implementation of advisory notices?

Management Summary:

- Top management should define and implement a process for continual improvement that can be applied to realization and support processes.
- Top management should create an environment that challenges people to seek opportunities for improvement of performance in processes, activities, and products.

8.5 Improvement

8.5.1 General

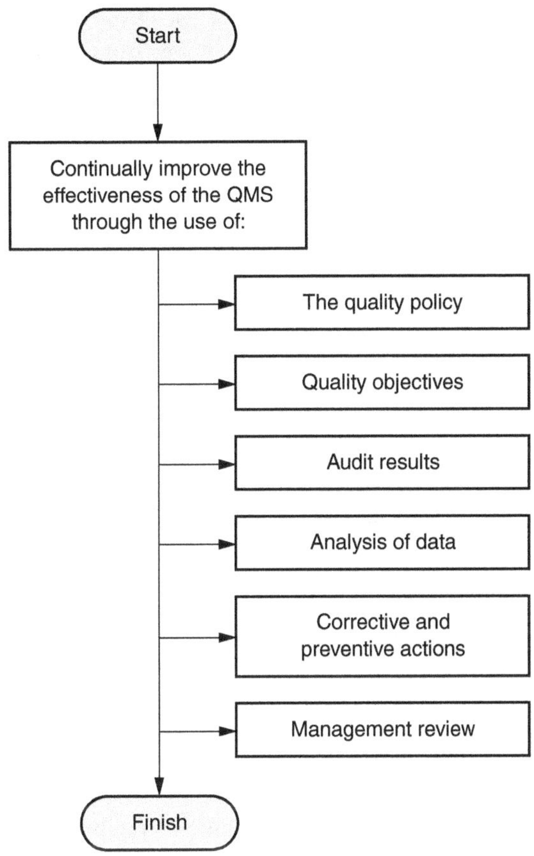

The Standard: 8.0 Measurement, Analysis and Improvement

8.5 Improvement

8.5.2 Corrective Action

13485	The organization must document the processes to be used to assess and correct nonconformities, eliminate their cause(s) to prevent future nonconformities from the same cause, and evaluate the effectiveness of the correction. Corrective action must be proportionate to the severity of the nonconformity, and documentation of corrective actions is to be treated as a quality record.
9004	Provides guidance on planning for corrective actions and determining causes. Offers 10 examples of information resources that help define effective corrective actions.
9000	Defines *corrective action* in 3.6.5 as "action taken to eliminate the cause of a detected nonconformity or other undesirable situation" and notes that it is meant to prevent recurrence of nonconformity, while *preventive action* is meant to avoid the possibility of a nonconformity ever occurring.

Document Requirements:

- Documented procedure
- Record (e)

Internal Audit Questions:

- Is there a written procedure that addresses corrective actions (or corrective action/preventive action [CAPA])?
- Does the organization take corrective action to eliminate causes of nonconformities?
- Is corrective action taken appropriate to the impact of the problems encountered?
- Is risk management (for product and the process) properly addressed?
- Do corrective action procedures provide for identifying nonconformities, determining causes, evaluating need for actions to prevent recurrence, determining the corrective action needed, and implementation of the needed corrective action?
- Are corrective actions taken verified to have no adverse effect on the product or process?
- Do corrective action procedures provide for recording the results of corrective actions taken?
- Do the corrective action procedures require that the effectiveness of the corrective actions implemented be assessed?

Management Summary:

- Corrective action should be used as a tool for improvement.
- Corrective actions should be included in management review.
- The organization should incorporate thorough impact analysis, risk assessment, and root cause analysis into the corrective action process.
- The effect and effectiveness of the actions taken must be assessed.
- Corrective action considerations include:
  - Customer complaints
  - Nonconformity reports
  - Internal audit reports
  - Inspection results/trending
  - Management review outputs

8.5 Improvement

8.5.2 Corrective Action

Corrective actions eliminate the causes of nonconformities to prevent recurrence.

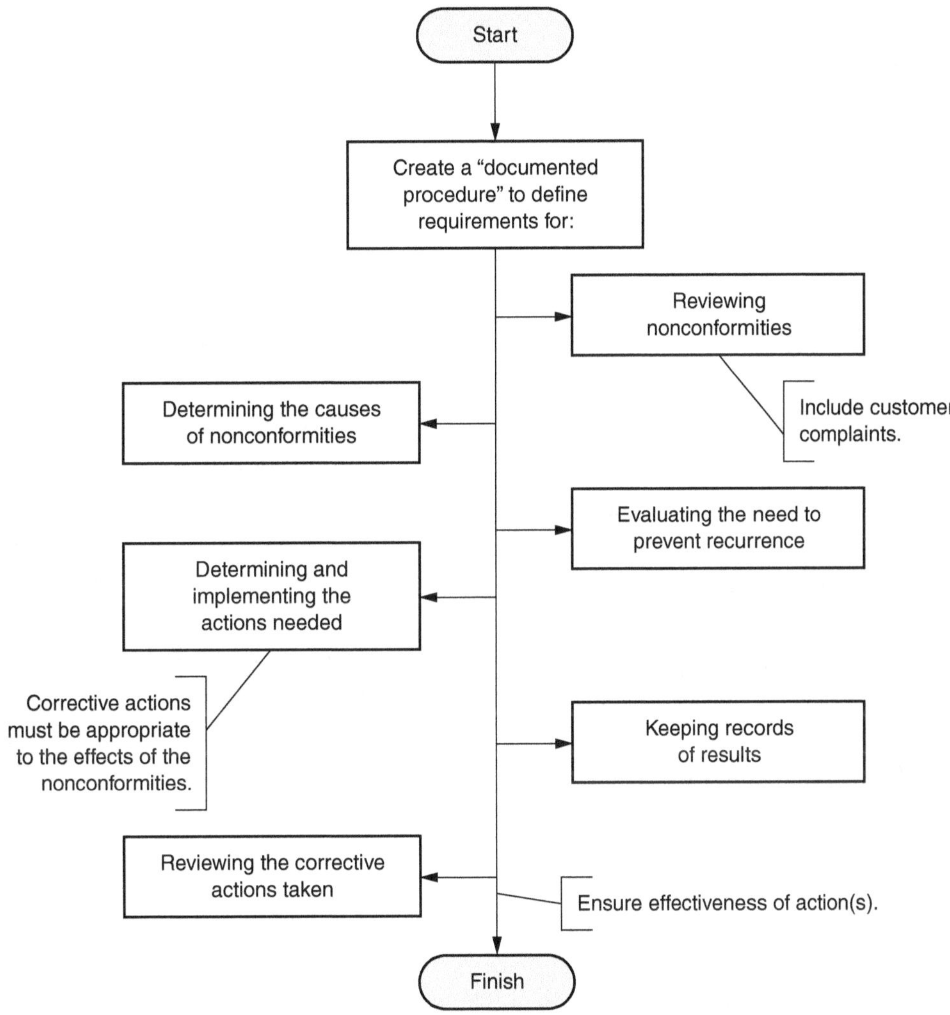

The Standard: 8.0 Measurement, Analysis and Improvement

8.5 Improvement

8.5.3 Preventive Action

13485	The organization must document the processes to be used to identify and eliminate potential causes to prevent nonconformities from occurring, with action proportionate to the severity of a potential nonconformity, and evaluate the effectiveness of the actions. Documentation of preventive actions is to be treated as quality records.
9004	Recommends planning for loss prevention and offers 12 sources for quantitative data to be used in the planning process.

Document Requirements:

- Documented procedure
- Record (d)

Internal Audit Questions:

- Is there a written procedure that addresses preventive actions (or CAPA)?
- Does the organization identify preventive actions needed to eliminate the potential causes of possible nonconformities?
- Is preventive action taken appropriate to the impact of potential problems?
- Do the preventive action documented procedures provide for identifying potential nonconformities and their probable causes?
- Do the documented procedures for preventive action provide for determining the need for preventive action and implementation of the preventive action needed?
- Are actions taken verified to have no adverse effect on the product or process?
- Do the preventive action documented procedures provide for recording the results of the preventive actions taken?
- Do the preventive action procedures require that the effectiveness of the corrective actions implemented be assessed?

Management Summary:

- The organization should develop a plan for loss prevention and apply it to realization and support processes, activities, and products.
- Results of the evaluation of the loss prevention plan should be considered as an output for management review and should be used in the improvement processes.
- The effect and effectiveness of the actions taken must be assessed.
- After verifying that CAPA does not have an adverse effect, actions should be taken without undue delay.

8.5 Improvement

8.5.3 Preventive Action

Preventive actions eliminate the causes of nonconformaties before they occur.

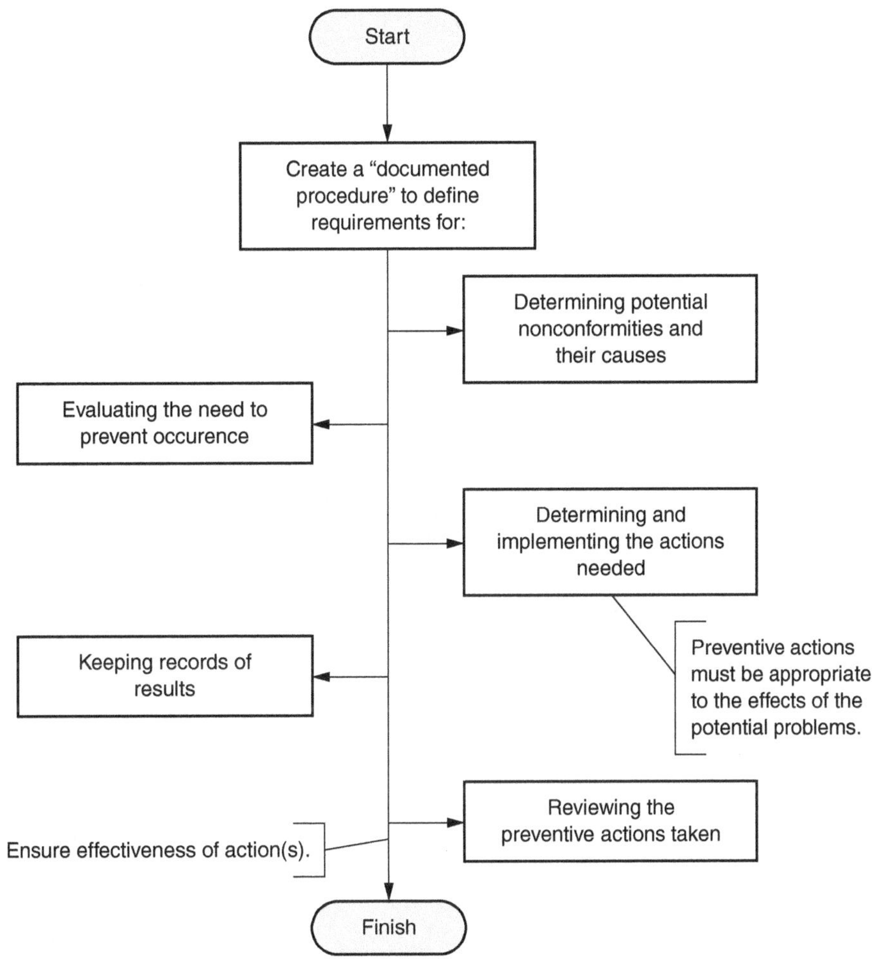

Clause	Change	Interpretation
8.2.1 (Feedback)	There is an emphasis on documenting all aspects of feedback to ensure that regulatory requirements are met. Postproduction activities are now to be included as part of feedback, which itself must now form an input into the risk management process.	Organizations are now required to document all aspects of feedback, which must now include feedback sources that may not be initially captured through the client's internal processes, for example, verbal feedback (phone). Organizations must now formalize postproduction activities as part of feedback, for example, adverse events, complaint handling within the QMS.
8.2.2 (Complaint handling)	Complaint handling is specifically called out in ISO 13485:2016, and requires clients to document the following: • Timely complaint handling that meets regulatory requirements. • Documenting, receiving, and recording information and performing evaluations to determine whether feedback is a complaint. • Determining the need to report to regulatory authorities. • Determining the need to initiate corrections and corrective actions.	Organizations may already be performing the requirements called out in this new clause, based on the previous revision of this standard; however, the new clause 8.2.2 contains specific requirements for complaint handling. This new clause again features the link to meeting regulatory requirements.
8.2.3 (Reporting to regulatory authorities)	There is emphasis that complaints that meet regulatory reporting criteria (adverse events, issuance of advisory notices) must now be documented. The process of notifying the regulatory authorities and records of these notifications must now be documented.	Organizations are now required to formally document their regulatory reporting process and criteria. This should also include the notifications that are sent, that is, adverse events, advisory notices, and so on.
8.2.6 (Monitoring and measurement of product)	Any records used to document data generated from measuring or test equipment must also have the identity of the equipment used recorded.	Organizations are now required to document the identification of the measuring or test equipment that is used to determine product requirements. This also provides traceability to the test equipment and links to its validation or measurement systems analysis (MSA) records.

Continued

Clause	Change	Interpretation
8.3.1 General	Controls and roles and responsibilities for identification, documentation, segregation, evaluation, and disposition of nonconforming product must now be defined. Evaluation of the nonconforming product must include determination of the need for an investigation and notification of any external party, actions taken, and the rationale for any decisions made.	Organizations must determine the roles and responsibilities of those involved in the control of nonconforming product. This clause specifies the revised requirements that need to be included in the client's procedure for control of nonconforming product.
8.3.2 (Actions in response to nonconforming product detected before delivery)	Nonconforming product can only be accepted under concession if a justification is provided and approval is obtained. Records of acceptance by concession, including the identity of the person authorizing the concessions, are maintained.	Organizations can use testing of product samples according to a sampling regime to assess the impact of the nonconformance on the fit, form, or function of the product. An organization's acceptance of concession product must be fully documented.
8.3.3 (Actions in response to nonconforming product detected after delivery)	This new clause is specific to managing nonconforming product after delivery. Procedures for issuance of advisory notices in accordance with regulatory requirements must be documented. This process must be able to be initiated at any time an adverse action or advisory notice is required. All records of actions relating to advisory notices must be maintained.	Organizations may already have a field action or recall procedure that covers these topics. Organizations should also use these procedures in conjunction with vigilance and risk management procedure(s) that feed into the advisory notice process to determine the severity, risk, and occurrence of the event.
8.3.4 (Rework)	Any rework performed on product shall be verified to ensure that it meets the original acceptance criteria and regulatory requirements. Organizations must ensure that records of rework activities are maintained.	Organizations must have specific requirements to ensure that any rework performed is tested against the original acceptance criteria and that the rework also meets regulatory requirements. Records of these activities must be maintained.
8.5 (Improvement)	When organizations make changes to their processes or product(s) to ensure medical device safety and performance, key inputs must be used, for example, vigilance, PMS (post-market surveillance).	Organizations must now use data sources that can impact the safety and performance of the device as inputs into the improvement process. PMS feedback is now included in this section as an input from the market.

SUMMARY

by S. Pletzer

ISO 13485:2016—8 Measurement, Analysis and Improvement

8.1 General

Plans and implements the monitoring, measurement, analysis, and improvement processes needed to meet product requirements, and to ensure the conformity and effectiveness of the QMS.

8.2 Monitoring and measurement

Measures customer feedback, conducts internal audits, monitors and measures processes and product characteristics.

8.3 Control of nonconforming product

Ensures that nonconforming product is identified and controlled, eliminates nonconformity sources, authorizes nonconforming product use under concession, or reworks product to meet original requirements.

8.4 Analysis of data

Determines, collects, and analyzes data to demonstrate suitability and effectiveness of the quality management system and make an evaluation of any needed improvements.

8.5 Improvement

Makes changes ensuring suitability and effectiveness of the quality management system through customer complaints, corrective action, and preventive action.

ISO 13485:2016 DOCUMENTED REQUIREMENTS FROM MOST STRATEGIC TO MOST TACTICAL

Section	Description	Documentation
5.3	Quality policy and quality objectives	Statements
4.2.2	Quality manual	Manual
4.2.4	Control of documents	Procedure
4.2.5	Control of records	Procedure
7.1 (b)	Product realization planning	Procedure
7.4.1	Purchasing process	Procedure
7.5.1	Production control	Procedure
7.5.5	Preservation of product (shelf life)	Procedure
7.5.6	Validation of computer software	Procedure
7.5.7	Validation of sterile processes	Procedure
7.5.8	Product identification	Procedure
7.5.8.1	Returned product identification	Procedure
7.5.9	Product traceability	Procedure
7.5.11	Preservation of product (processing)	Procedure
7.6	Control of monitoring and measuring equipment	Procedure
8.2.1	Feedback system	Procedure
8.2.4	Internal audits	Procedure
8.3	Control of nonconforming product	Procedure

Continued

Section	Description	Documentation
8.3.3	Issue and implementation of advisory notices	Procedure
8.4	Analysis of data	Procedure
8.5.2	Corrective action	Procedure
8.5.3	Preventive action	Procedure
4.2.4	Document control	Record
5.6.1	Management review	Record
6.2 (e)	Competence, awareness, and training	Record
6.3	Maintenance	Record
7.1 (b and d)	Planning of product realization	Record
7.1	Risk management	Record
7.2.2	Review of requirements related to product	Record
7.3.3	Design and development inputs	Record
7.3.4	Design and development outputs	Record
7.3.5	Design and development review	Record
7.3.6	Design and development verification	Record
7.3.7	Design and development validation	Record
7.3.8	Design and development transfer	Record
7.3.9	Control of design and development changes	Record
7.4.1	Supplier evaluations	Record
7.4.2	Purchasing information	Record
7.4.3	Verification of purchased product	Record
7.5.1	Batch records	Record
7.5.3	Installation activities	Record
7.5.4	Servicing activities	Record
7.5.5	Sterilization parameters	Record
7.5.6	Validation (process)	Record
7.5.7	Validation (sterilization)	Record
7.5.9.1	Product traceability	Record
7.5.9.2	Consignee name and address	Record
7.5.10	Customer property	Record

Continued

Section	Description	Documentation
7.5.11	Preservation of product	Record
7.6	Calibration (as found)	Record
7.6	Calibration (corrected)	Record
7.6	Computer software application	Record
8.2.2	Complaint handling (investigations)	Record
8.2.2	Customer complaints (no action justification)	Record
8.2.3	Regulatory reporting	Record
8.2.4	Internal audit	Record
8.2.6	Product inspection and testing	Record
8.3.1	Nonconformances (nature)	Record
8.3.2	Nonconformances (authorization)	Record
8.3.3	Nonconformances (after delivery)	Record
8.3.4	Rework	Record
8.4	Analysis of data	Record

EXECUTIVE SUMMARY

by S. Pletzer

ISO 13485:2016—Buy-In

- Management commitment/cross-functional commitment
- Internal commitment/cultural change

Top Management Involvement

- Leading by example
- Establishing and imparting a vision of the future
- Setting top-level policy and objectives
- Creating an environment that encourages involvement and development of people
- Identifying the processes that add value
- Direct participation in improvement projects

Identify Key Differences in ISO 13485:2016

- Understand process-based approach
- Measurable objectives
- Personnel competencies
- Effectiveness of actions
- Stronger supplier qualification and monitoring
- More management involvement
- Evaluation of monitoring/measurement results
- Meet customer's requirements
- Maintain effectiveness of QMS

ISO 13485:2016—Maintenance

- Management reviews
- Internal audits (first-party audit)
- Third-party audits

- Periodic document reviews/revisions
- Establish measurable objectives—you can't manage what you can't measure

Summary

- Utilizes process-based approach
- Top management commitment required
- Effectiveness of training
- Competencies of employees
- Communication

SOURCES OF INFORMATION

STANDARDS

- ISO 9000:2015 *Quality management systems—Fundamentals and vocabulary*
- ISO 13485:2016 *Medical devices—Quality management systems—Requirements for regulatory purposes*
- ISO 9004:2018 *Quality management—Quality of an organization—Guidance to achieve sustained success*

INFORMATIONAL WEBSITES

- http://www.iso.org/iso/home.html
- http://www.asq.org
- http://www.moorhill.com
- *Design Control Guidance for Medical Device Manufacturers.* Text related to FDA 21 CFR 820.30 and subclause 4.4 of ISO 9001
- *Labeling for Medical Devices.* Guidance on medical device labeling from the FDA
- *Quality Systems Regulations (QSR).* US FDA 21 CFR Parts 808, 812, and 820 Medical Devices; Current Good Manufacturing Practice (CGMP); Final Rule
- *Post-Market Surveillance.* Guidance on criteria and approaches for medical device post-market surveillance
- *Human Factors in Medical Device Design.* Discusses the CDRH push to have manufacturers apply human factors to device design
- http://www.fda.gov/

ADDITIONAL RESOURCES

- ISO 13485:2016 (ISO requirements). https://www.iso.org/standard/59752.html
- 21 CFR Part 820 (FDA regulations). https://www.accessdata.fda.gov/scripts/cdrh/cfdocs/cfcfr/CFRsearch.cfm?CFRPart=820
- *Medical Devices and Diagnostic Industry* (MD+DI) (Newsletter). https://www.mddionline.com
- Medical Design and Manufacturing (MD&M) (Trade shows). https://www.facebook.com/mdmtradeshows/

www.ingramcontent.com/pod-product-compliance
Lightning Source LLC
LaVergne TN
LVHW060038210926
848481LV00013B/3900
* 9 7 8 1 6 3 6 9 4 1 3 2 5 *